W9-ATD-586

Oxford Physics Series

Oxford Physics Series

1. F.N.H. ROBINSON: *Electromagnetism*
2. G. LANCASTER: *D.c. and a.c. circuits*
3. D.J.E. INGRAM: *Radiation and quantum physics*
5. B.R. JENNINGS and V.J. MORRIS: *Atoms in contact*
6. G.K.T. CONN: *Atoms and their structure*
7. R.L.F. BOYD: *Space physics; the study of plasmas in space*
8. J.L. MARTIN: *Basic quantum mechanics*
9. H.M. ROSENBERG: *The solid state*
10. J.G. TAYLOR: *Special relativity*
11. M. PRUTTON: *Surface physics*
12. G.A. JONES: *The properties of nuclei*
13. E.J. BURGE: *Atomic nuclei and their particles*
14. W.T. WELFORD: *Optics*
15. M. ROWAN-ROBINSON: *Cosmology*
16. D.A. FRASER: *The physics of semiconductor devices*

E.J. BURGE

PROFESSOR OF PHYSICS, CHELSEA COLLEGE
UNIVERSITY OF LONDON

Atomic nuclei and their particles

Clarendon Press. Oxford. 1977

Oxford University Press, Walton Street, Oxford OX2 6DP

OXFORD LONDON GLASGOW NEW YORK
TORONTO MELBOURNE WELLINGTON CAPE TOWN
IBADAN NAIROBI DAR ES SALAAM LUSAKA ADDIS ABABA
KUALA LUMPUR SINGAPORE JAKARTA HONG KONG TOKYO
DELHI BOMBAY CALCUTTA MADRAS KARACHI

Casebound ISBN 0 19 851834 X
Paperback ISBN 0 19 851835 8

Printed in Great Britain
by Thomson Litho Ltd., East Kilbride, Scotland

Editor's Foreword

Nuclear physics, and especially the particles of nuclear physics, have always provided a fascination for sixth formers, undergraduates, and research workers alike. It is a subject which is not only of inherent interest but also forms part of the basic framework of any course in modern physics. It is for this reason that the Oxford Physics Series have included this as one of their core texts. Taken with a parallel volume on ***Atoms and their structure***, they follow on from ***Radiation and quantum physics*** and lead on to ***The properties of nuclei***

This volume therefore stands at the borderline between first and second year undergraduate work, but the introductory chapter provides a very readable background for the student first entering a physics course and also for the interested sixth former. The text is written with a careful balance between ideas on the pure science side, such as those on symmetry and the conservation laws, and the implications on the applied side, as exemplified by fission and fusion. In a text of limited length many topics have necessarily received rather brief treatment, but readers will find that the topics treated reach to our current boundaries of knowledge, and reference forward to more detailed work is included.

D.J.E.I.

Preface

Nuclear physics is both popular and important. Its popularity is evident in schools and continues right through to postgraduate studies. Its importance is evident not only in the emphasis on fundamental research with very high energy machines, but also in numerous applications. Most of these, ranging from nuclear power stations to the use of nuclear radiation in medicine, need to be seen in relation to the environment. Research into elementary particles continues to produce remarkable results that deepen our understanding of the forces of nature and show how limited are our 'commonsense' conceptions of space and time. It is not surprising that self-respecting scientists and engineers recognize that they require a basic understanding of nuclear physics.

This volume is intended to be equally suitable for the university entrant specializing in physics and for the chemist, geologist, engineer, or biologist who is taking a course covering basic science. The mathematics required is not very demanding, and the supporting physics is not too difficult for non-specialists, except for a few sections that could be omitted.

The introductory chapters are largely historical, partly in order to present the ideas required to follow the discovery of the nucleus and the explosive development of the subject, and partly to allay the prevalent tendency to divorce physics as a specialized discipline (at times almost an ideology) from physicists, other scientists, and society in general. Later chapters are increasingly arranged in conceptual or logical order, and become more demanding. Many new concepts need to be introduced in order to bring the reader near enough to the frontier of the subject to appreciate some of the excitement and interest surrounding contemporary discoveries. Inevitably, some topics are treated very briefly, and simple derivations are used in the hope of illuminating the subject without unduly misleading the reader. The problems at the end of each chapter include many that develop points from the text, and are of a fair range of difficulty.

The author remembers with gratitude the late Professor G.K.T. Conn, who made many helpful comments, especially on the first three chapters, and jointly considered the relation of this text to his work on ***Atoms and their structure*** in this Series. The other directly related text in this Series, ***The properties of nuclei*** by Dr G.A. Jones, assumes familiarity with many of the topics treated here, including accelerators and detectors, and Dr Jones has kindly read this text and made valuable suggestions. The author is grateful to the several friends and colleagues who have taken an interest in parts of the text and contributed to the minimization of errors and the improvement of clarity and style. Finally to Dr D.J.E. Ingram is due much appreciation for his editing of the script and his encouraging remarks, and to Mrs J Ashdown many thanks for her careful typing.

ACKNOWLEDGEMENTS

The author wishes to thank the American Physical Society and the authors concerned for permission to reproduce or adapt material for Figs. 3.5, 3.6, 5.6, and 8.3.

E.J.B.

Contents

1. ATOMS, ELECTRONS, AND PHOTONS 1

Introduction. Atoms. Electrons. Early theories of atomic structure. Photons. Problems.

2. NUCLEAR PHYSICS BEFORE THE NUCLEUS 15

Radioactivity. Types of radiation - alpha, beta, gamma. Positive ions. Mass-energy relation. Problems.

3. THE NUCLEUS REVEALED BY ALPHA PARTICLES 36

Introduction. Alpha-particle scattering. Differential cross-section. Rutherford's theory. Diffraction effects in alpha scattering. Alpha-particle spectra. Alpha emission. Alpha-particle reactions. Problems.

4. NUCLEAR ACCELERATORS 64

Introduction. Continuous acceleration. Repeated acceleration. Intersecting storage rings. Problems.

5. NUCLEAR INSTRUMENTS AND METHODS 79

Energy loss by nuclear particles. Particle detectors. Track detectors. Isotopes. Problems.

6. NUCLEAR REACTIONS 100

Introduction. Energetics. Cross-sections. Neutron reactions. Fission and chain reactions. Fusion reactions. Problems.

CONTENTS

7. NUCLEAR FORCES AND MODELS 121

Introduction. The deuteron. Neutron-proton scattering. Yukawa's field quantum - the pion. The weak interaction - beta emission. Summary of types of interactions. Nuclear models. Problems.

8. COSMIC RAYS AND ELEMENTARY PARTICLES 143

Introduction. Cosmic rays. Positrons and antiparticles. Muons and lepton numbers. Hadrons. Particle resonances. Classification of hadrons. Conservation laws and symmetry. Alpha and omega and beyond - the discovery of charm. Problems.

APPENDIX A. RUTHERFORD SCATTERING THEORY 175

APPENDIX B. TRANSMISSION THROUGH A POTENTIAL BARRIER 180

APPENDIX C. LINEAR-MOMENTUM CONSERVATION AND LINEAR-TRANSLATION INVARIANCE 184

ANSWERS TO NUMERICAL PROBLEMS 187

INDEX 189

PHYSICAL CONSTANTS AND CONVERSION FACTORS 194

1. Atoms, electrons, and photons

INTRODUCTION

Solids and liquids are readily recognized by the senses, but gases and vapours cannot be appreciated so directly. Leaves rustle and trees bend, and so we talk of the wind. By similar processes of thought we talk of magnetic fields when the compass needle is returned to a stable direction. The attraction of small objects by substances such as glass 'electrified' by friction introduces the idea of electric fields. As our experience widens, our powers of conceptualization increase, and so does our ability to manipulate these concepts by appropriate mathematical techniques, notably the differential and integral calculus and, more recently, numerical methods using computers.

From time to time crucial observations are made or striking syntheses of ideas are propounded, and in retrospect they become landmarks of experience. We may mention only a few: Newton's understanding of gravitation (1687) linking the terrestrial to the celestial; Maxwell's electromagnetic theory (1864) linking electricity, magnetism and light; Einstein's theory of relativity (1905) linking energy and mass and identifying the importance of frames of reference. It would be tempting for some to include Dalton's atomic theory (1803), but that would be an injustice to the very long history of development of the concepts of atoms and molecules. The naming of the electron as the smallest electric charge in electrolysis was the suggestion of Johnstone Stoney (1894) but the identification of the definitive properties of the corresponding particle is generally attributed to J. J. Thomson (1897). The introduction of the quantum or photon, on the other hand, is without question the work of Planck (1900), but it was not much appreciated until Einstein (1905) used the photon to explain the photoelectric effect. Planck's work has as its counterpart the work of de Broglie (1924) linking the wave and particle aspects of matter, which initiated the development of quantum mechanics.

The identification of the nucleus of the atom by Rutherford (1911) is one of the great landmarks in science. It was followed by the recognition of two companions to the gravitational interaction and the electromagnetic interaction, namely, the strong nuclear interaction and the weak interaction. The strong nuclear interaction is involved in the binding of protons and neutrons to form the nucleus, and in the scattering of neutrons, by protons (see pages 123 ff). The weak interaction is involved in beta decay, i.e. the emission of high-energy electrons from nuclei, and related phenomena (see pages 21 ff, and page 130). The study of the nucleus also led directly to the development of nuclear power for both military and peaceful purposes, and to the many applications of radioactivity in medicine, industry, and research.

We shall review in Chapter 2 the events that prepared the way for the discovery of the nucleus, and then consider in Chapter 3 the importance of alpha-particle studies in extending our understanding of nuclear properties. The rapid advances in nuclear physics went hand-in-hand with the development of nuclear accelerators and particle and track detectors, which are described in Chapters 4 and 5. Among the wealth of possible nuclear reactions we emphasize fission and fusion (Chapter 6) for their importance in power generation, and in Chapter 7 the fundamental strong nuclear interaction between neutron and proton is treated, together with a brief account of beta emission and nuclear models. Finally, in Chapter 8, an introduction is given to cosmic rays and the fascinations and problems of elementary particles, which are believed to provide clues to the understanding of nuclear forces and to the fundamental symmetries underlying experiments and theories.

In this chapter we shall consider briefly the history of the concepts of atoms, electrons, and photons. Thereby the reader will be able to link familiar concepts and experimental methods to the new material to be presented. The scene will also be set for the experiments and ideas, outlined in Chapter 2, which preceded the discovery of the nucleus.

ATOMS

The concept of the atom can be traced back to the Greeks, Leucippus and Democritus in the sixth and fifth centuries B.C. respectively, but it was the work of the eighteenth and nineteenth centuries A.D. which led to its definition as 'the smallest particle of matter that retains the chemical properties of an element'. The word 'chemical' is important, and indicates the context of much of the work that led Dalton in 1803 to postulate atoms. He was thereby able to explain the law of conservation of mass in chemical reactions (Lavoisier 1789) and the law of definite proportions (Proust 1799), and he extended the latter to the law of multiple proportions (1804). Dalton's own words ('New system of philosophy', 1808) are transparently clear. 'If there are two bodies, A and B, which are disposed to combine, the following is the order in which combinations may take place, beginning with the most simple: namely, 1 atom of A + 1 atom of B ..., 1 atom of A + 2 atoms of B ..., 2 atoms of A + 1 atom of B, etc.' We might wish to emphasize 'may take place' since not all that many multiples are in practice 'disposed to combine'.

Significantly, Dalton wrote in 1803 'An inquiry into the relative weights of the ultimate particles of bodies is a subject, so far as I know, entirely new. I have lately been prosecuting this enquiry with remarkable success'. Although it was necessary for the accuracy of these measurements to be improved, and the distinction between atomic weights, molecular weights, and combining weights or equivalent weights had yet to be made, there is no doubt that Dalton's work was a major advance. The ordering of elements by their atomic weights about 1860 led to the discovery of periodicities in their physical and chemical properties. In 1871 Mendeléev was reasonably successful in systematizing the known elements in his periodic table, mainly because he was bold enough both to leave gaps for undiscovered elements, and also to make small reversals of order in the belief that some weights were inaccurate.

From relative weights of atoms we turn to their sizes. A number of estimates based on the kinetic theory of the viscosity of a gas and the density of the corresponding liquid were reviewed by Kelvin in 1870, and led to values of about 10^{-10} m for the diameter of an atom. As early as 1865 Loschmidt used such estimates of size to deduce the Avogadro constant, i.e. the number of molecules in the gram molecular weight, now known to be 6.023×10^{23} mol^{-1}. From this number we can deduce the masses of individual atoms, ranging from 1.67×10^{-27} kg for hydrogen to 395×10^{-27} kg for uranium.

With sizes and masses confidently measured the validity of the atomic theory of matter was beyond serious challenge, although a number of influential scientists (e.g. Ostwald and Mach) opposed the theory even into the twentieth century. The concept of the atom as indivisible, however, was soon to need revision, as a result of experiments on electrical discharges in gases.

ELECTRONS

Electrostatic attraction phenomena using amber and fur were reported by the Greeks about 600 B.C. William Gilbert in the sixteenth century studied such phenomena and coined the word 'electricity' from elektron, the Greek word for amber. In 1733 two kinds of electrification were noted by DuFay, and in 1747 Benjamin Franklin independently introduced the terms 'positive' and 'negative' electricity.

The use of gases at reduced pressures for electrical investigations dates back to 1705, when Hawksbee noticed that light was emitted in the frictional electrification of amber in a partially evacuated vessel. By about 1870 the study of low-pressure discharges between metal electrodes with controlled potential differences was beginning to produce significant results. The associated glowing of the walls of the glass discharge tubes was traced to the impact of 'rays' from the cathode, i.e. the negative electrode. William Crookes and other British workers suggested that the rays consisted of high-velocity negatively charged particles, but Hertz, Lenard, and others favoured an interpretation in terms of short-wavelength electromagnetic waves. Perrin in 1895 collected the particles in a cylinder connected to an electrometer, and demonstrated their negative charge.

The successes of Thomson in 1897 were directly due to the development of better vacuum systems. He was able to deflect the particles with both electric and magnetic fields, whereas electric fields used by earlier workers had had no detectable effect, since the cathode rays produced many ions in the residual gas of the relatively poor vacua and these acted as an electric shield. In one of a comprehensive series of experiments, Thomson measured the velocity v of the particles of charge e by crossed electric (X) and magnetic (B) fields (Fig. 1.1) acting over the same length of the particle beam and adjusted to give zero deflection; then $eX = evB$, and hence $v = X/B$. The velocity produced by this accelerating voltage was about 3×10^7 m s^{-1}, i.e. about one tenth of the velocity of light, by far the highest velocity ever measured by man up to that time, other than the velocity of light itself.

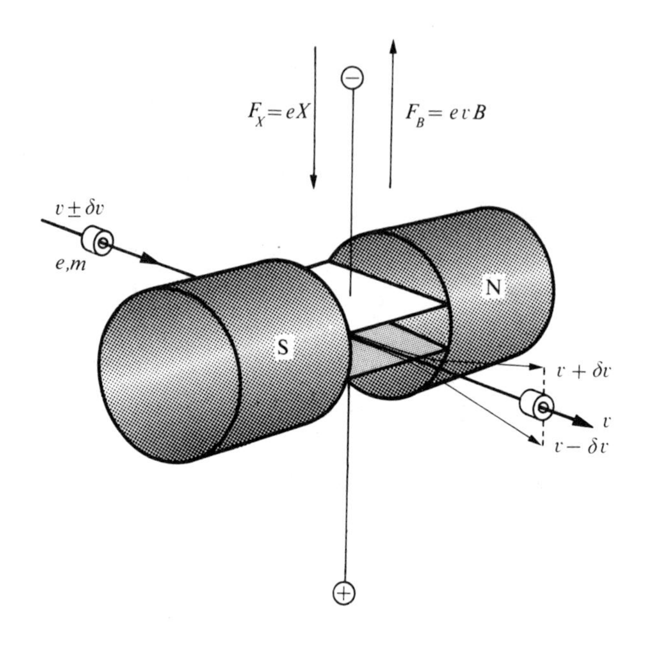

FIG. 1.1. Crossed electric and magnetic fields acting as a velocity selector (Wien filter) for charged particles.

Thomson used two independent methods to measure the ratio of the mass m to the electric charge e of the particles. In the first he measured the total charge Ne = Q carried by N particles and also their total energy $\frac{1}{2}Nmv^2 = W$. The magnetic rigidity, i.e. magnetic field B times the radius R of the deflected beam of particles, was then measured and is given by BR = mv/e (since for the circular path $evB = mv^2/R$). Then $m/e = B^2R^2Q/2W$. In the second method he used the technique of crossed electric and magnetic fields to determine v, and then measured the angular deflection θ (Fig. 1.2) caused solely by the electric field X acting over the length ℓ of the beam. The acceleration

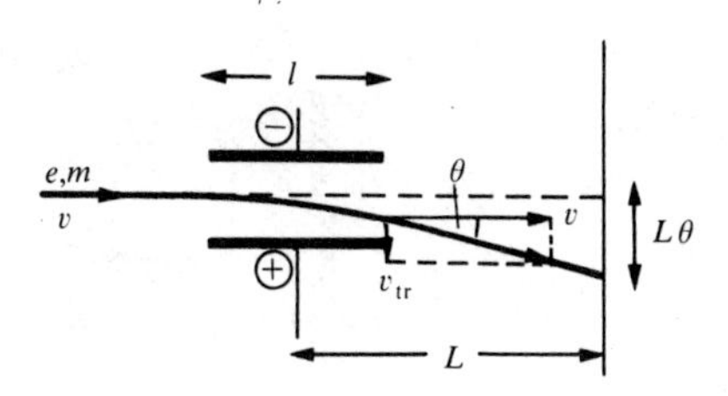

FIG. 1.2. Deflection of a charged particle beam of known velocity by an electric field.

perpendicular to v due to X is given by eX/m and acts for a time ℓ/v, producing a transverse velocity at exit from the field given by $v_{tr} = eX\ell/mv$. Since θ is small it is equal to the ratio of this velocity to the initial velocity of the beam, hence $\theta = eX\ell/mv^2$. Inserting $v = X/B$ yields $m/e = \ell B^2/X\theta$.

The values obtained for m/e were essentially independent of the nature of the residual gas in the tube (air, hydrogen, carbon dioxide) and of the material of the electrodes (aluminium, platinum) and were about one thousandth of the value for the hydrogen ion as found from electrolysis experiments. Similar results were obtained about the same time by Kaufmann and Wiechert. At first Thomson wrote 'the smallness of m/e is, I think, due to the largeness of e [compared with that carried by an ion in an electrolyte] as well as the smallness of m.'

It remained for Thomson to measure the charge on the individual particles, but the cathode rays did not lend themselves to the methods available. He therefore first showed the identity of cathode rays and photoelectrons by measuring m/e for the latter. Then he used a technique developed by C. T. R. Wilson (later to be of great value in the nuclear track detector) known as the cloud chamber, in which ions in a gas act as centres for condensation of supersaturated water vapour. The radius of the drops formed can be estimated from the downward velocity of the cloud. The total weight of the cloud and the charge it carries then give the charge per droplet. The value was evidently equal in magnitude to the (positive) charge carried by the hydrogen ion in electrolysis. The work of Millikan (1909) based on his well-known oil-drop method improved considerably the accuracy of the measurement of the electronic charge. Measurements by Bäcklin (1928) based on X-ray diffraction data and a knowledge of the Avogadro constant N_A were even more precise, and they served to reveal a systematic error in Millikan's value arising from the use of an incorrect value of the viscosity of air. The accepted value of e is -1.60×10^{-19}C. The mass of the electron was estimated by Thomson to be about 1.4×10^{-3} of that of the hydrogen ion m_p. The accepted value of m_e is 9.1×10^{-31} kg, i.e. about $1/1836 \times m_p$, and together with the value of e can be regarded as the characteristic parameters of the electron.

EARLY THEORIES OF ATOMIC STRUCTURE

Since neutral atoms in a gas under the action of ultraviolet light, or X-rays, produced electrons and positive ions, it was evident that the atom can be divided, in spite of the usual interpretation of its name. The structure of the atom became a matter for speculation, and Thomson in 1899 suggested that there are 'a large number of corpuscles [i.e. electrons]' in a positively charged space. By 1904 he had developed the idea in terms of '[electrons] in a series ... of concentric shells ... moving about in a sphere of uniform positive electrification', and he explicitly related this to the periodic classification of the elements, an idea first put forward in his 1897 paper on the electron. Measurements of the scattering and absorption of X-rays by gases were interpreted by Thomson, and indicated that the number of electrons per atom was of the order of the atomic weight - in fact it is equal to the atomic number - and therefore most of the mass was in the positively charged part of the atom.

The next significant development in theories of atomic structure was the result of Rutherford's interpretation of alpha-particle scattering in terms of the nucleus (1911), but it is worth noting, in the same volume of The Philosophical Magazine that contained Thomson's 1904 paper, the neglected publication of the Japanese physicist Nagaoka. Nagaoka compared the atom to the planet Saturn and continued '[the atom] will evidently be approximately realized if we replace these satellites [i.e. Saturn's rings] by negative electrons and the attracting centre by a positively charged particle.' He attempted to explain characteristic atomic spectra, and even radioactivity, but states clearly 'the objection to such a system of electrons is that the system must ultimately come to rest, in consequence of the exhaustion of energy by radiation, if the loss be not properly compensated.'

PHOTONS

The suggestion that light consists of particles which enter the eye from luminous or illuminated bodies was current in the circle of Pythagoras about 500 B.C. Newton revived the possibility in the seventeenth century but his writings allow that undulatory or oscillatory features are likely to be present. His contemporaries, Hooke and Huygens, favoured a wave theory, and of course the well-known Newton's rings experiment finds a ready explanation in terms of periodic phenomena or concepts. The explanations of (1) diffraction effects at the boundaries of shadows, (2) interference as, for example, between light from nearby slits illuminated by a common source, and (3) polarization effects were given by Young and by Fresnel in terms of wave theory early in the nineteenth century. Visible light was shown to have a wavelength of about 5×10^{-7} m, i.e. about 1/2000 th of a millimetre, and together with the velocity of light (3×10^8 m s^{-1}) this yields a frequency of about 10^{15} Hz (hertz = cycles per sec). Colours were explained by different wavelengths - longer for red ($\simeq 7 \times 10^{-7}$ m) and shorter for blue ($\simeq 4 \times 10^{-7}$ m).

Faraday established a relation between magnetism and electricity, and suspected that there was a relation between electromagnetism and light. He succeeded in 1845 in detecting a rotation of the plane of polarization of a beam of light passing through lead glass by applying a magnetic field parallel to the direction of the beam. This is now known as the Faraday effect. However, it was not until 1865 when Maxwell developed his four electromagnetic equations, including the new concept of displacement current, that the possibility arose of interpreting light as an electromagnetic wave. Maxwell's equations showed that a plane-polarized wave in vacuo consists of electric and magnetic fields having directions perpendicular both to each other and to the direction of propagation of the wave. For monochromatic light these fields vary in magnitude at any one point in space as $\sin \omega t$, where ω is the angular frequency ($\simeq 10^{15}$ Hz for visible light), and at any one time the fields vary as $\sin(2\pi x/\lambda)$ along the beam direction x, where λ is the wavelength. Their velocity is given by $c = (\varepsilon_0 \mu_0)^{-\frac{1}{2}}$, where μ_0 is the magnetic permeability of a vacuum ($= 4\pi \times 10^{-7}$ H m^{-1} exactly by definition) and ε_0 is the electric permittivity of a vacuum ($= 8.854 \times 10^{-12}$ F m^{-1}). Electric and magnetic measurements confirmed the accuracy of these quantitative predictions, and the experiments of Hertz (1887) with oscillatory electric discharges clearly demonstrated the existence of electromagnetic waves with wavelengths of the order of a metre, i.e. short radio waves. The elegance, accuracy, and apparent completeness of Maxwell's theory seemed at first to leave little opportunity for developments in the fundamental understanding of the propagation of light and other electromagnetic energy.

There were, however, clouds on the horizon. Maxwell's equations, unlike the laws of dynamics, change their form when the classical space and time transformations are applied for Galilean (i.e. 'commonsense') frames of reference. But careful experiments had shown that electromagnetic processes are observed to be independent of the (Galilean) frame of the observer. This was one of the problems that required the theory of relativity presaged by Lorentz and propounded by Einstein. Again, there was the unsolved problem of the wavelength distribution of the radiation emitted by a hot 'black' body, as measured, for example, by Lummer and Pringsheim (1897). Rayleigh and Jeans applied classical concepts and Maxwell-Boltzmann statistics to the emitting oscillators of the body and obtained agreement for the longest wavelengths, but predicted infinite emission at the smallest wavelengths. (For a more detailed discussion of this, and of subsequent developments, see Radiation and quantum physics by D. J. E. Ingram in this series.)

In 1900 Planck broke away from the continuity of thought of classical physics in his explanation of black-body radiation. He first obtained an accurate empirical formula and then recognized the necessary assumption, namely, that the oscillators of frequency ν could emit energy only in 'energy elements', now known as quanta, of magnitudes $h\nu$, $2h\nu$, ..., $nh\nu$, where n is a whole number and h is a constant, the Planck constant, 6.626×10^{-34} J s. Even for light with $\nu \simeq 10^{15}$ Hz, $h\nu$ is a minute amount of energy - for green light it is about 2.5 eV, i.e. the kinetic energy of an electron accelerated from rest across a potential difference of 2.5 V. When it radiated, the oscillator was considered to radiate all of its energy $nh\nu$, but each oscillator was assumed to absorb energy in a continuous manner, and the radiation field around the body was not assumed to be populated with quanta. The observed form of the variation of intensity per unit wavelength interval with wavelength was

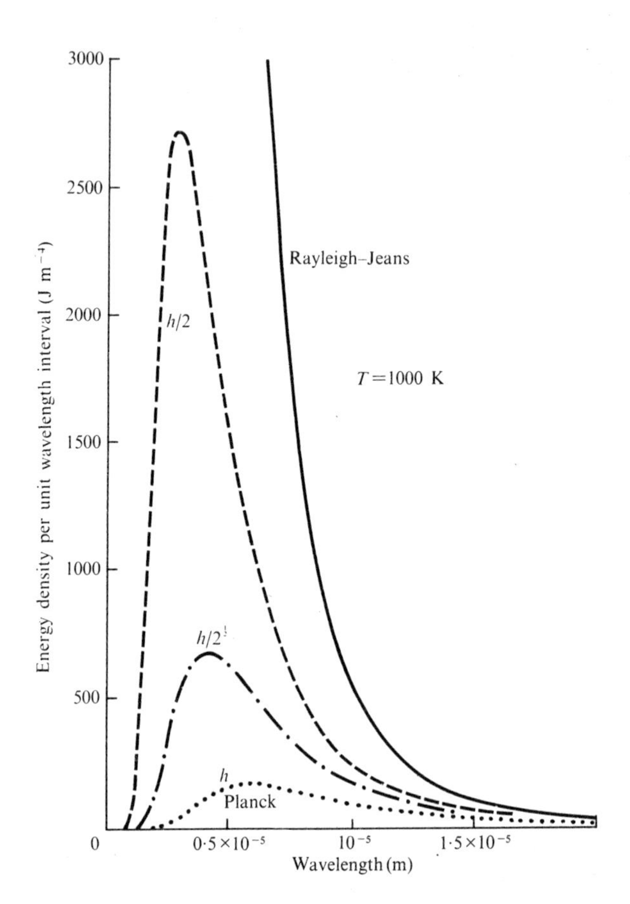

FIG. 1.3. Relation between classical and quantum theories of black body radiation. As $h \rightarrow h/2^{\frac{1}{2}} \rightarrow h/2 \rightarrow 0$, Planck's expression transforms to that of Rayleigh and Jeans.

accurately fitted by Planck's equation simply by finding the appropriate magnitude of h (Fig. 1.3). Notice that for values of h approaching zero, Planck's formula reduces to the classical values of Rayleigh and Jeans.

The photoelectric effect (a by-product of the experiments of Hertz (1887) and first studied by Hallwachs (1888)) is the emission of electrons from matter when it is illuminated by light or other electromagnetic radiation of adequately short wavelength. Einstein (1905) interpreted the threshold wavelength λ_{th} as that of a photon which has just sufficient energy $h\nu_{th} = hc/\lambda_{th}$ to remove from the material the electron with the smallest binding energy $e\varphi$, usually referred to as the work function. For ultraviolet light $h\nu$ is about 5 to 10 eV, and any photon of energy greater than $h\nu_{th}$ ($= e\varphi$) can yield photoelectrons with kinetic energies up to a maximum value $T_e(\max) = h\nu - e\varphi$. The success of this equation quickly established the need to consider electromagnetic radiation in terms of photons whenever the energy involved in the process is of the order of $h\nu$.

For many people, it is a great temptation to try to form a mental picture of the photon. Commonly, but erroneously, the photon is conceived as being extremely small in size. It then comes as a surprise to learn that interference phenomena can be recorded and built up, photon by photon, for two slits several metres apart. The question arises, does each photon sample both slits? Another common suggestion is that light is emitted and absorbed in quanta yet propagated as waves, but in order to investigate its propagation it is necessary to interact with the light and it interacts ultimately in terms of quanta. Wave phenomena require for their interpretation a knowledge of the phase of the wave, and the accuracy with which this phase can be expressed or measured increases with the number of photons involved. The phase is essentially indeterminate for one photon.

It must be recognized, however, that the lack of sophistication and caution in the mental pictures entertained by many pioneering physicists did not prevent, and might even have facilitated, their making very important experimental and theoretical advances.

PROBLEMS

1.1. The velocity of electrons in one of J. J. Thomson's experiments in 1897 was found to be 3×10^7 m s^{-1}. What can be deduced about the accelerating potential difference that he used?

1.2. There are five interrelated parameters in the design of Thomson's apparatus depicted in Figs. 1.1 and 1.2, namely, θ, v, ℓ, X, and B. Suggest suitable values for these quantities. Comment on the expected accuracy in θ, bearing in mind the size of the electron beam. Hence consider the accuracies of v, ℓ, X, and B, and deduce the expected accuracy in e/m.

1.3. A candle flame can be seen at night at a distance of 2 km by the unaided eye. If it emits visible light at the rate of 0.02 J s^{-1}, what is the approximate number of photons entering the eye per second? State clearly your assumptions and approximations.

1.4. What is the wavelength of a beam of ultraviolet light if its photon energy is 10 eV? If the beam produces photoelectrons from iron, which has a work function of 4.72 eV, what is the maximum energy and the maximum velocity of the electrons?

2. Nuclear physics before the nucleus

RADIOACTIVITY

Nuclear physics was born in 1896, ten years before Rutherford fathered the nucleus. The occasion was the accidental discovery of radioactivity by Becquerel, during experiments on the X-rays found by Roentgen in the previous year. In this chapter we consider how these X-ray experiments led to the identification of alpha, beta, and gamma rays, and how the laws of radioactive decay were formulated and the radioactive series explored. The properties of alpha and beta particles are then considered, together with the characteristics of their emission. Finally, we introduce positive ions and isotopes, and the important relation between mass and energy. This chapter, therefore, sets the scene for the discovery of the nucleus, presented in Chapter 3.

In 1895 Roentgen noticed that electrons in cathode-ray tubes made the glass walls luminesce, i.e. produce visible light, but some compounds such as zinc sulphide and barium platinocyanide gave a much more intense effect than glass. He also discovered that a discharge tube totally enclosed in black cardboard still produced a strong luminescence on a nearby piece of paper coated with barium platinocyanide. Roentgen identified the penetrating rays as X-rays, 'for the sake of brevity', and noted that they also fogged photographic plates. Their nature as electromagnetic waves of very short wavelengths (about 10^{-10} m) was established in 1912, by Friedrich and Knipping, using a crystal as a diffraction grating.

Henri Becquerel (1896) attempted to find a relation between X-ray luminescence and the 'fluorescence' in potassium uranyl sulphate, i.e. the production of light of a particular wavelength by the salt during exposure to light of a shorter wavelength, e.g. sunlight. A photographic plate was wrapped in black paper, and a thin crystal of uranium salt placed on it. Exposure to sunlight and subsequent development revealed local darkening, as if X-rays had been produced by the sun's action! Three sunless days in February 1896 proved to be a blessing in disguise, for it was found that prepared plates and crystals stowed unexposed in a drawer showed remarkable blackening when developed. It was evident that rays, rather like X-rays in their action, were emitted by the uranium crystals, without the stimulus of the sun's rays and, in Becquerel's words, 'the time of persistence is infinitely greater than that of the visible radiations emitted by such bodies.'

The investigation of 'radioactivity' - the word was coined by Marie Curie in 1898 - was pursued in several countries. Some of the 'rays' were readily stopped by 0.02 mm of aluminium, and Rutherford called these alpha rays. They were later shown to be doubly ionized helium atoms. A second type required a stopping thickness of about 1 mm of aluminium and were termed 'beta-rays', soon to be identified by Becquerel (1900) as fast electrons because of their negative charge and e/m value.

A third type of radiation was recognized by Villard in 1900, and came to be known as gamma rays in 1903. These rays were not deflected by a magnetic field and had considerable penetrating power. Satisfactory evidence of their electromagnetic nature was not obtained until 1914 when Rutherford and Andrade showed that gamma rays could be diffracted by crystals, and had wavelengths shorter than X-rays.

After the three sunless days of February 1896, Becquerel concentrated his attention on the new radiation from uranium and soon showed that it was independent both of the state of chemical combination of the element and of its temperature. Once it was realized that the rays readily ionized gases, since they could discharge an electroscope, new tools for their study became available. Marie Curie and her husband Pierre Curie entered the scene in 1898, and quickly showed that thorium had similar radioactive properties. From pitchblende they separated a new substance 'whose activity is about 400 times as great as that of uranium' - 'we propose to call it polonium' (after Marie's homeland, Poland). An even more active element, radium, was later identified by the Curies, and confirmed by enormously laborious chemical separations that yielded 0.1 g of radium chloride from one ton of pitchblende residues.

The time-dependence of radioactivity was conjectured by Becquerel after Crookes had shown in 1900 that the active substance in uranium nitrate could be chemically separated from the uranium. Becquerel argued that since all natural uranium salts had a characteristic 'equilibrium' activity proportional to the uranium present, the chemically 'deactivated' uranium would recover its equilibrium activity - and it did! But Becquerel had to wait eighteen months for what he described as 'complete recovery'.

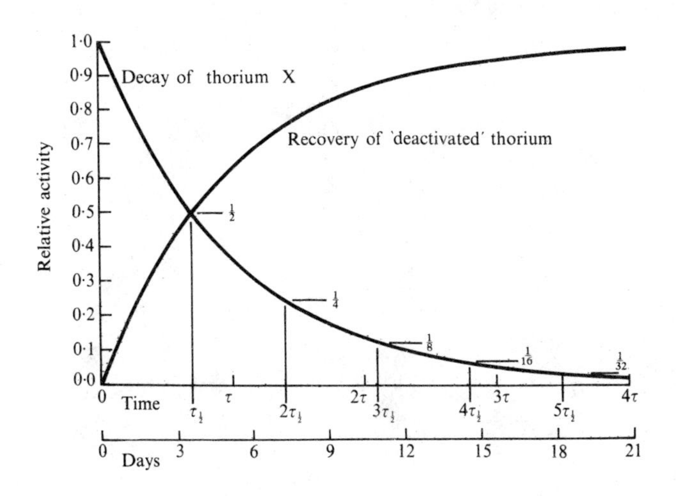

FIG. 2.1. Decay of the chemically-separated active component of thorium, and recovery of the 'deactivated' thorium.

Rutherford and Soddy (1902), working with thorium under rather special conditions (see Problem 2.4), obtained results similar to those shown in Fig. 2.1 (on elevation to the peerage, Rutherford had a version of this diagram incorporated in his escutcheon). Thorium X was the active component chemically separated from the natural thorium, and its decay curve had as its complement the curve of recovery of the deactivated thorium. It was shown that the reactivated thorium can once again have its thorium X activity separated and will again recover. The explanations soon followed:

1. Emission of alpha particles by thorium is accompanied by a change of element, i.e. there is a transmutation of the chemical element - the dream of the alchemists come true.
2. The element formed, thorium X, is itself strongly radioactive, and can be separated chemically from thorium.
3. Thorium X decays to a very weakly radioactive product.
4. Natural thorium robbed of thorium X decays to produce more thorium X until equilibrium is attained, i.e. until the rate of production of thorium X equals its inherent decay rate.

The variation of the activity A of thorium X with time is a simple exponential,

$$A = A_o \exp(-\lambda t) \quad ,$$

where λ is the radioactive constant of thorium X and A_o is the activity at time $t = 0$. This may be derived from the basic assumption that λ is the probability per unit time that a given nucleus of thorium X will decay. For N nuclei at time t,

$$dN/dt = -\lambda N.$$

Integration of $dN/N = -\lambda dt$ yields $\ln N = -\lambda t + \text{constant}$. At $t = 0$, $N = N_o$ and the constant is $\ln N_o$ hence $N = N_o \exp(-\lambda t)$. But the activity A, i.e. the number of disintegrations per second, is λN, and $\lambda N = \lambda N_o \exp(-\lambda t)$ or $A = A_o \exp(-\lambda t)$. For thorium X, $\lambda = 2.20 \times 10^{-6}\ s^{-1}$. A more useful characteristic constant is the 'half-life' $\tau_{\frac{1}{2}}$, which is simply the time for N_o nuclei of the species to decay to $\frac{1}{2}N_o$. Since $N_o/2 = N_o \exp(-\lambda\tau_{\frac{1}{2}})$,

$$\tau_{\frac{1}{2}} = (\ln 2)/\lambda = 0.693/\lambda.$$

For thorium X, $\tau_{\frac{1}{2}} = 3.64$ days (Fig. 2.1).

The form of the recovery of activity by the deactivated thorium is $A_2 = A_\infty\{1-\exp(-\lambda t)\}$ and follows readily from the assumptions:

Thorium	→	Thorium X	→	Product
$\lambda = \lambda_1$		$\lambda = \lambda_2$		$\lambda \to 0$
$N_o = N_1(0)$		$N_o = 0$		$\tau_{\frac{1}{2}} \to \infty$
$N_t = N_1$		$N_t = N_2$		

At any time t, the rate of production of thorium X = $\lambda_1 N_1$
and the rate of decay of thorium X = $\lambda_2 N_2 = A_2$.
Hence

$$dN_2/dt = \lambda_1 N_1 - \lambda_2 N_2.$$

Since thorium decays very slowly (half-life = 1.39×10^{10} years) its activity $A_1 = \lambda_1 N_1(0) \exp(-\lambda_1 t)$ is essentially constant during the period of the experiment. Then $dN_2/(A_1 - \lambda_2 N_2) = dt$, or $\ln(A_1 - \lambda_2 N_2) = -\lambda_2 t + \text{constant}$; at $t = 0$, $N_2 = 0$ and therefore the constant = $\ln A_1$. It follows that

$$A_2 = \lambda_2 N_2 = A_1 \{1 - \exp(-\lambda_2 t)\}.$$

The commonsense interpretation of this result is evident since the fractions, had they not been separated, would together have had a constant activity A_1 ($=A_\infty$), i.e.

$$A_1 = A_1 \exp(-\lambda_2 t) + A_1 \{1 - \exp(-\lambda_2 t)\}$$

natural thorium = thorium X + deactivated thorium

The activity is measured in curies, defined in 1948 as 1 curie = 1 Ci = 3.7×10^{10} disintegrations per second - a value very close to the activity of the radon found in radioactive equilibrium with one gram of radium, which was the basis of the original definition in 1910. Radon is a gaseous radioactive element, $_{86}Rn$. Recently a new unit of activity was approved: 1 becquerel = 1 Bq = 1 disintegration per sec.

It often happens in physics that simplifying assumptions lead to important advances in understanding. Rutherford and Soddy realized that more than three components were involved in the thorium decay scheme, but the simple model of thorium, thorium X, and final product works well. The detailed picture revealed by later studies appears at first so very different that it provides a useful problem (see Problem 2.4) for the reader to explain why the early experiments produced the results they did, and what reformulation is required of the simple assumptions of only three components. In Fig. 2.2 the full sequence of radioactive changes is given for the series that contains thorium. Three radioactive series are found in nature, characterized by their atomic mass numbers (1) A = 4n (thorium), (2) A = 4n + 2 (uranium), and (3) A = 4n + 3 (actinium). Each ends with a stable isotope of lead. The fourth series with A = 4n + 1 is not found in nature but has been produced artificially - the neptunium series (see Chapter 6).

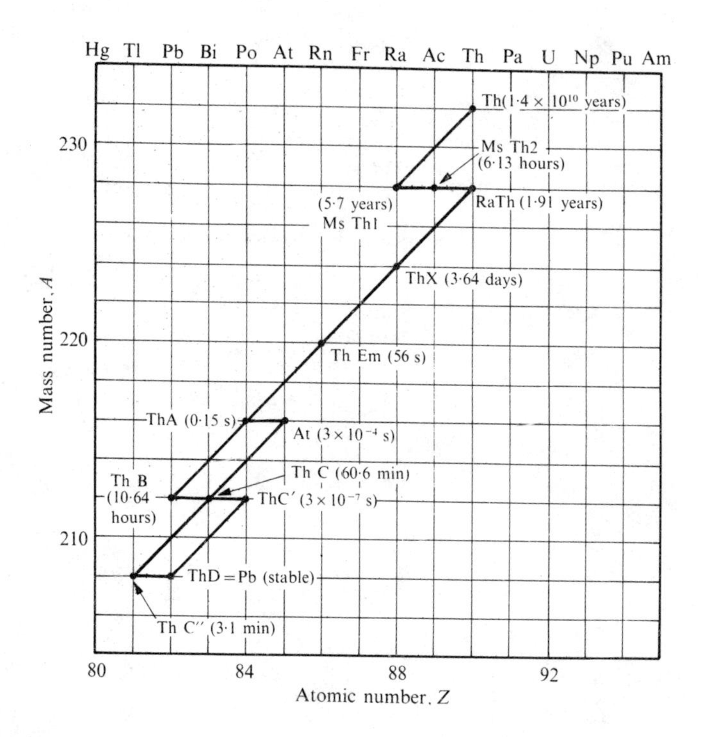

FIG. 2.2. The thorium radioactive series. Diagonal transitions correspond to alpha-emission ($\delta A = -4$, $\delta Z = -2$) and horizontal transitions correspond to (negative) beta-emission ($\delta A = 0$, $\delta Z = +1$).

TYPES OF RADIATION - ALPHA, BETA, GAMMA

The identification of alpha rays was not such an easy matter as the pin-pointing of beta rays by their e/m value. Deflections of alpha particles by magnetic and electric fields yielded a ratio of positive charge to mass which was about one half of that of the hydrogen ion. In 1908 Rutherford and Geiger found the charge to be +2e, strongly suggesting that the alpha particle is a doubly ionized helium atom of mass four times that of the hydrogen ion. Their experiment to determine the charge on the alpha particle was the first in which the Geiger counter (see pages 22 ff) was used, and deserves more attention.

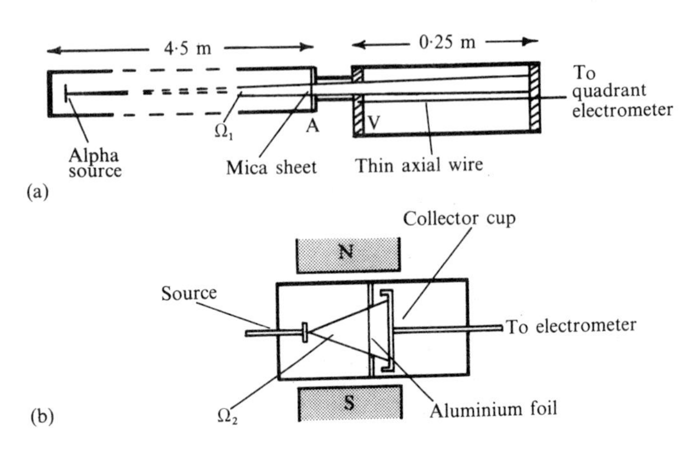

FIG. 2.3. Determination of the electric charge of the alpha-particle by Rutherford and Geiger. (a) Direct count, by Geiger counter, of rate for small solid angle. (b) Current flowing for large solid angle.

An alpha source in an evacuated enclosure (Fig. 2.3(a)) subtended a known solid angle Ω_1 at a small aperture A covered by a thin mica sheet, which allowed the particles to enter a cylindrical volume V at a pressure of a few torr. A thin axial wire was maintained at a potential of about 1300 V with respect to the cylinder, and the entry of a single particle produced a noticeable flick of the needle of the attached quadrant electrometer. The number n emitted into the small solid angle Ω_1 in a given time was therefore counted directly. For a much larger solid angle Ω_2 the charge Q in a given time was measured directly in a simple apparatus (Fig. 2.3(b)). A strong magnetic field perpendicular to the axis of the alpha beam prevented beta particles from the source reaching the shallow cup collector C, and prevented the escape from C of secondary electrons produced by the alpha bombardment. The aluminium foil stopped recoil atoms from the source. The charge q on each particle was then given by $Q\Omega_1/n\Omega_2$, and was unambiguously +2e.

An elegant experiment in confirmation of the identification of the alpha particle as a He^{2+} ion was performed by Sir James Dewar (1908) and repeated by Rutherford and Royds in 1909. Alpha rays from a source sealed by a thin window entered an evacuated glass vessel terminating in a capillary tube containing two electrodes. After a number of days the gas produced was compressed into the capillary and an electric discharge revealed the characteristic spectrum of helium gas.

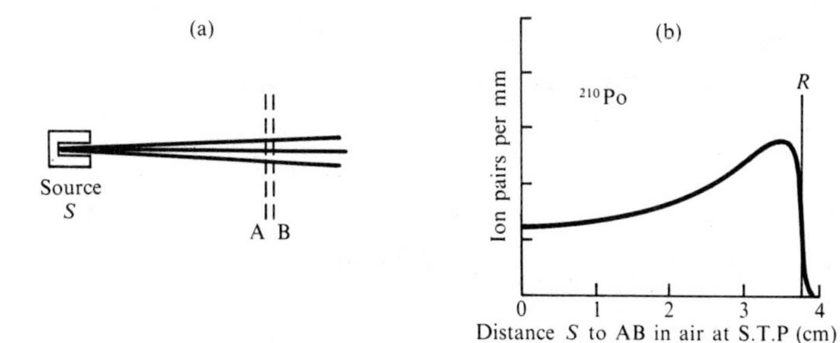

FIG. 2.4. Ionization produced by alpha-particles in air. (a) Thin ionization chamber AB. (b) Variation with distance from source, showing range R.

The ionization produced by alpha particles during their passage through a gas was investigated by W. H. Bragg (1904) using the arrangement shown schematically in Fig. 2.4(a). The ionization between the two grids, A,B, about 1 mm apart, was measured by the electric current flowing between A and B when a sufficiently large voltage was applied to collect all the ions initially produced in that region by the beam of alpha particles. An example of the values observed is shown in Fig. 2.4(b). As the alpha particles ionize the gas and lose energy the probability of ionization increases - later work showed that this probability is approximately inversely proportional to the particle energy.

Particles from a given source usually have a well-defined range R; some sources have several groups each with a characteristic range (see pages 50 ff). The small variation in the range of particles arises from the 'straggling' associated with the statistical fluctuations in the ionization processes.

Careful measurement of alpha-particle ranges and velocities led Geiger to the approximate relation T_α (in MeV) = $2.12\ R^{\frac{2}{3}}$ where T_α is the initial energy of the alpha particle and R is the range in centimetres of air at standard temperature and pressure (S.T.P.). Rutherford suggested in 1907 that there might be a relation between the range R of an alpha particle and the radioactive constant λ of the corresponding alpha emission. In 1911 Geiger and Nuttall proposed the rule, or relation,

$$\ln \lambda = A \ln R + B ,$$

where A and B are constants; using Geiger's range relation this can be rewritten

$$\ln \lambda = A_1 \ln T_\alpha + B_1 .$$

The interpretation of this relation by Gurney and Condon, and independently by Gamow, in 1928, was one of the first significant successes of wave mechanics in nuclear physics (see pages 53 ff and Appendix B).

Advances in the study of beta particles prior to Rutherford's discovery of the nucleus were less evident than those involving alpha particles. The identification of beta particles as fast electrons had been quickly established by Becquerel (1900) from their e/m value. For some years their energy spectrum was a matter of dispute - were they monoenergetic as suggested by von Baeyer, Hahn, and Meitner (1911), or with a range of energies, or both? A convincing demonstration of the continuous energy

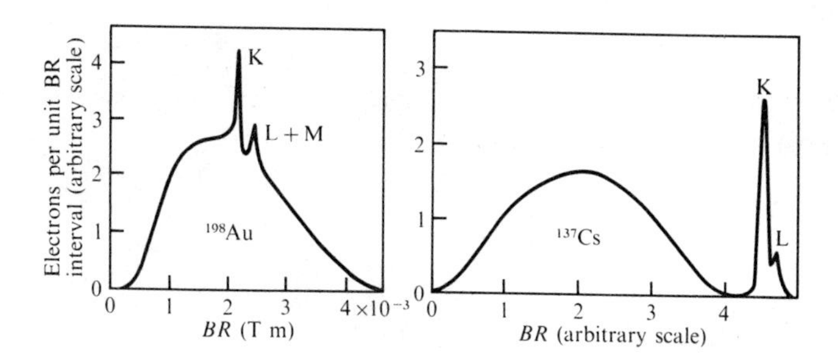

FIG. 2.5. Beta ray spectra with K, L and M conversion electrons. BR is the 'magnetic rigidity' of the electrons, i.e. the product of the applied magnetic flux density B in the magnetic beta-spectrometer and the radius R of the circular path of the electron. BR = p/e where p is the momentum of the electron.

spectrum of the beta particles emitted by nuclei was provided by Rutherford and Robinson in 1913, using a magnetic deflection method similar to that of Thomson. But there are also monoenergetic groups of electrons, for certain radioactive substances, with intensities that are usually only a few per cent of the total radioactivity (See Fig. 2.5 for recent results). These line spectra of electrons are found in association with not only the continuum of beta particles but also with the discrete alpha-particle spectra. They are now known to be the result of a direct transfer of (quantized) nuclear energy from an excited nucleus to one of the atomic electrons, leaving the nucleus in a lower excited state or in its ground state. Unfortunately they are commonly referred to as 'conversion electrons' because of the early suggestion by Rutherford (1914) that they are produced by gamma rays emitted from the excited nucleus which are converted via a photoelectric effect with the atomic electrons. This can happen, but is very improbable.

The continuous range of energies in the beta-particle

spectrum presented formidable problems. When a body of well-defined mass m_1 spontaneously breaks up into two particles each with definite mass m_2, m_3, the available kinetic energy T is shared in a unique way. This follows from the conservation laws of linear momentum and energy, thus

$$m_2v_2 = m_3v_3 = p \quad \text{and} \quad T = \tfrac{1}{2}m_2v_2^{\,2} + \tfrac{1}{2}m_3v_3^{\,2}$$

(in the non-relativistic case). Then

$$T = (p^2/2)\ (1/m_2 + 1/m_3) \quad \text{and} \quad T_2 = T\ m_3/(m_2 + m_3)\ ,$$

with $T_3 = T\ m_2/(m_2 + m_3)$. The range of energies for the beta particles therefore appears to rule out two-particle break up. Could there be two or more betas in each disintegration? This was ruled out by a number of considerations, including chemical identification of the initial and final nuclei which indicated a unit change of nuclear charge. Could there be a continuous range of energy states (and therefore masses) for the initial or final nucleus? That would imply a continuous range of gamma energies, and such a gamma spectrum from nuclei is never observed; all gamma spectra consist of discrete lines. Desperate attempts to solve these problems included a proposal that energy and momentum may not be conserved. In 1931 Fermi developed the suggestion by Pauli that a curious new particle, the neutrino, is emitted with the electron. This particle has zero rest mass, and therefore travels with the velocity of light; it is electrically neutral and extremely difficult to detect (pages 79 ff). The theory was strikingly successful in its explanation of the beta-particle spectrum, and the neutrino has now been detected by several distinct methods.

Very little was known about gamma rays in the first fifteen years after the discovery of radioactivity. It was noticed that they were always associated with alpha emission or beta emission, i.e. a pure gamma emitter was not found in nature. Once their energies could be measured in the mid-1920s the opportunity arose to correlate them with alpha-particle spectra (see pages 50 ff). Their interactions with matter led to new discoveries, e.g. the positron.

POSITIVE IONS

The discovery of the electron in gaseous discharges prompted a search for a corresponding positively charged particle. Goldstein in 1886 observed luminous rays repelled from the anode and passing through holes in the cathode. About the turn of the century, Wien measured the charge-to-mass ratio q/M for these particles by means of electric and magnetic deflections and the values were even less than that for hydrogen ions in solutions, suggesting particles thousands of times heavier than electrons. Wien soon showed that, assuming positive charges of the same magnitude as the electron charge, the masses corresponded to the atomic masses of the gas in the discharge. The lightest positive particle corresponded to the hydrogen atom or ion, and as late as 1914 Rutherford still described it as 'the positive electron', a terminology now reserved for the positron (see pages 144 ff). The name proton was not used explicitly for the hydrogen ion until 1920.

J. J. Thomson began to develop apparatus for more precise measurements of q/M for positive rays in 1907 and by 1911 was able to separate two distinct components in neon, one to be assigned a mass of approximately 20 and the other 22, on a scale with oxygen set at 16. His apparatus for e/m of the electron used electric and magnetic fields perpendicular to each other and to the electron beam. As shown in Fig. 2.6, for positive ions he used an electric field X which was parallel to a magnetic field B. It was shown on pages 6 and 7 that the angular deflection θ

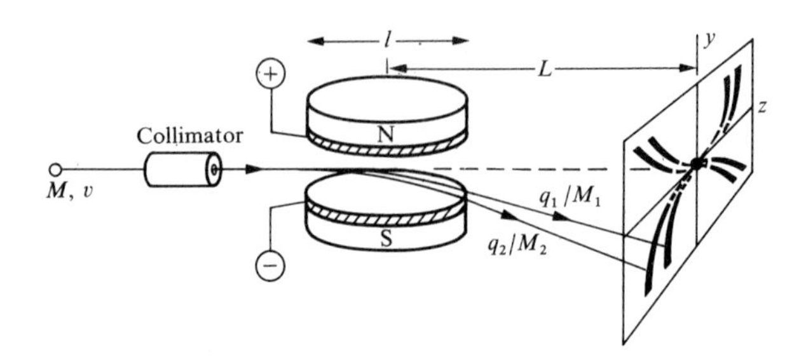

FIG. 2.6. Schematic view of J. J. Thomson's mass spectrograph.

produced by X alone is given by $\theta = qX\ell/mv^2$, where ℓ is the length of the (undeflected) beam over which X acts and v is the particle velocity. The corresponding displacement y on the screen, distant L from the centre of the field region, is simply $L\theta$ (Fig. 1.2) or $y = (X\ell L)(q/M)/v^2$.

The corresponding displacement z on the screen for the magnetic field B acting alone is given for small deflections by $(B\ell L)(q/M)/v$. This follows readily from Fig. 2.7, since $z \simeq L\varphi$, $\varphi \simeq \ell/R$ and R is obtained from $Mv^2/R = qvB$. It is clear that, since y is proportional to $1/v^2$ and z is proportional to $1/v$, the curve produced on the screen by both X and B acting together for ions with a range of velocities will be part of the parabola

$$z^2 = (\ell LB^2/X)(q/M)y ,$$

and there will be a separate curve for each value of q/M. Since the maximum velocity for a discharge tube with voltage V is given by $qV = \frac{1}{2}Mv^2$, the minimum value of y is $X\ell L/2V$.

Aston, working with Thomson, managed by means of gaseous diffusion to effect a partial separation in bulk of the two types of neon. In the same year (1913) Soddy collated the results of many workers and described radioactive species of identical

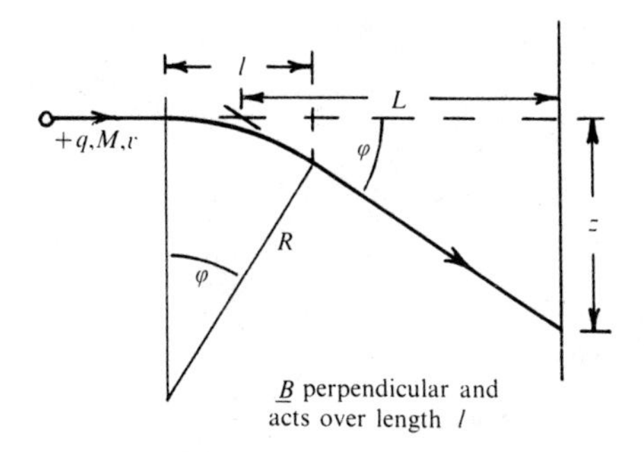

FIG. 2.7. Deflection of charged particle beam by a magnetic field.

chemical properties as isotopes, i.e. they occupied the same place (Greek: topos) in the periodic classification of the elements. Thereby was brought together the important realization that atoms of the same chemical element, containing the same number of electrons, could nevertheless have different masses, whether light as in the case of neon or heavy as found among radioactive elements. Precise measurements of isotopic masses are of considerable importance in studies of nuclear structure and reactions.

MASS-ENERGY RELATION

The history of most advances in science is a tapestry of interacting ideas. Energy was considered to be manifested in various forms in classical physics, the total amount in a closed system being conserved and its magnitude continuous. But in problems requiring quantum physics it is found to appear in discrete amounts, yet still in various forms and still conserved. By the early nineteenth century mass had come to be regarded as conserved, and in terms of atoms and molecules it was therefore treated as a definite, immutable property. When an atom of mass m_1 was split into an electron m_2 and a positive ion m_3 it seemed natural to assume that the total mass was still conserved, i.e. $m_1 = m_2 + m_3$. But it was later to be discovered that this equation is not exact.

Abraham (1903) and others made many attempts to account for the electron mass entirely in terms of the magnitudes of its associated electromagnetic field but they all failed. Lorentz applied to moving electrons the so-called Lorentz-Fitzgerald contraction, first proposed in 1893 in order to account for the Michelson-Morley experiment (1881-7) which had failed to detect movement of the Earth relative to the 'fixed ether'. If the mass of an electron with a uniform spherical charge distribution is taken to be entirely electromagnetic in origin, the mass will be inversely proportional to its radius; Lorentz obtained the relation $m = m_0(1 - v^2/c^2)^{-\frac{1}{2}}$ where m is the mass at velocity v, m_0 is its rest mass, and c is the speed of light. The argument used is now generally agreed to be fallacious, but the formula was correct and came at a convenient time to account for the results of Kaufmann (1901) and Bucherer (1908) on the variation with velocity of e/m for an electron. Kaufmann used a beam of beta particles in parallel electric and magnetic fields (cf. Thomson's positive-rays apparatus) to produce traces that departed from parabolas for the highest velocities ($\simeq$ 0.94c) at which the mass was $\simeq 3.1m_0$. Bucherer used a velocity filter with crossed electric and magnetic fields, and deflected the resultant monoenergetic beam with the same magnetic field acting alone. The elegant geometry allowed simultaneous measurement of a wide range of velocities, and in his work he used beta particles with velocities up to 0.68c. Numerous experiments and machines (e.g. nuclear accelerators) have since been based on the validity of $m = m_0(1 - v^2/c^2)^{-\frac{1}{2}}$ for any particle, charged or neutral, and thereby it has been tested to a high order of accuracy.

Whereas Lorentz viewed his derivation of the Fitzgerald

contraction in terms of purely formal space-time transformations, Einstein (1905) was bold enough to make the revolutionary suggestion that the Lorentz transformation expressed relations between <u>physical</u> space and time. The absolute frame of reference of the ether was abandoned, and the first postulate of special relativity emerged, namely, that physical observations and their corresponding theoretical interpretations (i.e. physical 'laws') are equally valid for all observers moving with constant linear velocities relative to each other. (General relativity, on the other hand, is concerned with accelerated frames of reference or observation.) The second postulate is that the velocity of light in vacuo is constant for all observers. The variation of mass with velocity follows directly from these postulates, and independently of whether the particle is charged or neutral, and of whether or not its mass is electromagnetic in origin. Another extremely significant derivation by Einstein relates the increase in mass to the increase in kinetic energy of the particle, or more generally $m_oc^2 + T = mc^2 = E$ where T is the kinetic energy and E is the total energy, i.e. the sum of T and the rest mass equivalent energy m_oc^2.

A simple derivation of $T = (m - m_o)c^2$ can be made starting from $m = m_o(1 - v^2/c^2)^{-\frac{1}{2}}$. The change in kinetic energy dT produced by a force F acting on a body moving through a distance dx is simply $dT = Fdx$. But force is the rate of change of momentum $F = d(mv)/dt$ and hence $dT = d(mv)dx/dt = vd(mv) = v^2dm + mvdv$. The relation $m = m_o(1 - v^2/c^2)^{-\frac{1}{2}}$ is conveniently re-expressed as $m^2(c^2 - v^2) = m_o^2c^2$, and differentiation yields

$$(c^2 - v^2)dm - mvdv = 0.$$

But

$$v^2\,dm + mvdv = dT\ ,$$

as shown earlier, hence $dT = dm.c^2$, and for a particle

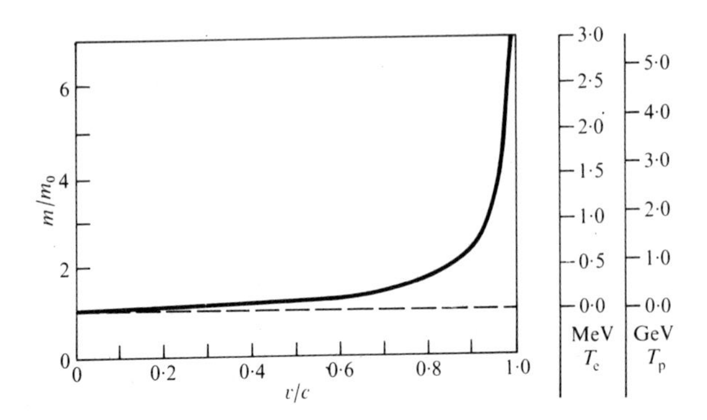

FIG. 2.8. Relativistic increase of mass with velocity (with magnitudes for electrons and protons).

accelerated from rest $T = (m - m_o)c^2$.

The magnitude of relativistic increases in mass with velocity, and the corresponding kinetic energies for electrons (in MeV, mega electron volts) and protons (in GeV, giga electron volts) can be seen in Fig. 2.8. It is evident that none of the particle velocities observed in physics before 1900 were large enough to produce detectable changes of mass. Furthermore, only electrons are readily accelerated to high enough velocities for the fractional change in mass to be measured - even so the potential differences required are hundreds of thousands of volts. Fortunately the energies of beta particles range up to one or two million electron volts and they enabled Kaufmann and Bucherer to measure the mass variation.

The equivalence of mass and energy is not restricted to kinetic energy. If energy Q needs to be injected into a system (e.g. an atom) to split it into two parts (e.g. an electron and a positive ion) then in the simplest case where the two parts are separated, but at rest, the sum of their rest masses $m_2 + m_3$ will be equal to the rest mass of the atom m_1 plus the mass equivalent

Q/c^2 of the injected energy, i.e. $m_1 + Q/c^2 = m_2 + m_3$. The energy Q is referred to as the binding energy of particle 2 with particle 3. For atoms the energy to remove the most loosely bound electron, i.e. the first ionization potential (Fig. 7.5(a), page 134), is in the range ~ 2 eV to ~ 25 eV (e.g. it is 13.6 eV for hydrogen). Now the mass equivalent of 13.6 eV is 2.4×10^{-35} kg, and it is clear that such small changes in mass cannot be detected by any normal weighing techniques. The very latest methods of mass spectrometry, developed from techniques similar to those of Thomson, can just detect the mass equivalents of electron binding energies. The energy required to remove a proton or a neutron (Fig. 7.5(b)) from a nucleus is approximately one million times the atomic electron binding energies, and its measurement is relatively easy. Such nuclear binding energies are directly involved in the production of power from nuclei, whether by fission or fusion, and it is not surprising that Einstein's equation $E = mc^2$ is so frequently associated in many people's minds with nuclear physics.

PROBLEMS

2.1. Marie and Pierre Curie remarked that for their first sample of polonium the 'activity is about 400 times as great as that of uranium', and for radium it is even greater. Comment on these statements in the light of later measurements of the half-lives: $\tau_{\frac{1}{2}}(^{238}\text{U}) = 4.51 \times 10^9$ years, $\tau_{\frac{1}{2}}(^{210}\text{Po}) = 140$ days, $\tau_{\frac{1}{2}}(^{226}\text{Ra}) = 1620$ years.

2.2. How 'complete' was the recovery of Becquerel's 'deactivated' uranium after his eighteen-month wait? The half-life of uranium X may be taken as 24.1 days.

2.3. Explain in words the expression for the mean life τ of a radioactive nucleus in a sample of N_0 such nuclei of radioactive constant λ,

$$\tau = \frac{1}{N_o}\int_o^\infty t(\lambda N.dt).$$

Hence substitute $N = N_o \exp(-\lambda t)$ and integrate by parts to show that $\tau = 1/\lambda$. (Lifetimes of unstable elementary particles, discussed in Chapters 7 and 8, are expressed as 'mean life' rather than 'half-life.' See Table 8.1.)

2.4. Study Fig. 2.2 and discuss expected observable differences between the simple thorium-thorium X scheme proposed by Rutherford and the series of decays now known to take place. (Hint: the thorium was in a thin layer, and a steady flow of air was maintained through the electroscope and the container.)

2.5. One gram of radium in radioactive equilibrium with its decay products emits 1.3×10^{11} alpha particles per second.
(a) Calculate the volume of helium gas at S.T.P. produced in a year. (b) Sir James Dewar used 70 mg of radium chloride ($RaCl_2$) in radioactive equilibrium and detected helium after waiting 1000 hours. What volume of gas at S.T.P. did he manage to detect? (Atomic weights: Ra 226, Cl 35.5.)

2.6. The presentday abundancies of ^{238}U and ^{235}U are in the ratio 137.8:1. Estimate the age of the Earth, assuming that the abundance ratio was originally 1:1.
$\tau_{\frac{1}{2}}(^{238}U) = 4.5 \times 10^9$ years, $\tau_{\frac{1}{2}}(^{235}U) = 7.13 \times 10^8$ years.

2.7. The radioactive constants for the members of the radioactive series $A \rightarrow B \rightarrow C \rightarrow D$ are 4λ, 3λ, 2λ and λ respectively, and the decay product of D is stable. Show (a) that the total activity of the sample decreases as $\exp(-\lambda t)$ and (b) that this condition is independent of the initial proportions of A, B, C, and D.

2.8. (a) A mixed beam of protons, deuterons, and alpha particles accelerated from rest by a potential difference of 90 kV moves perpendicular to a uniform magnetic field of flux density 1.0 T. Calculate the radii of the paths of the three types of particle. (Assume ratios of masses to be 1:2:4.) (b) What would happen if the same beam entered an electric field of 500 kV m^{-1}, in a direction perpendicular to this field, between parallel electrodes bent to form concentric semicircles of radii such that the path of the proton beam is also a (concentric) semicircle?

2.9. In J. J. Thomson's positive-ray apparatus there are five design parameters to be selected X, B, ℓ, L and the accelerating potential difference on the gaseous discharge tube. Suggest suitable values and plot the expected parabolas for singly charged neon ions of mass numbers 20 and 22.

2.10. Expand $(1 - v^2/c^2)^{\frac{1}{2}}$ to show that for low velocities the kinetic energy T of a particle of rest mass m_o is given by $T = \frac{1}{2}m_o v^2$. At what value of $T/m_o c^2$ will this classical formula be in error by 10 per cent, and what is the corresponding value of T for electrons? Does the classical formula give a value of $T/m_o c^2$ that is too large or too small?

3. The nucleus revealed by alpha particles

INTRODUCTION

The discoveries of one generation often become the tools of the next. There was a very short interval between the discovery of X-rays and their application in medical diagnosis and research, the result, in part, of the stimulation of popular imagination by newspaper reports. In similar fashion, the medical aspects of radioactivity were investigated soon after Becquerel in 1901 suffered skin burns as a consequence of carrying a radium source in his pocket for a few days. Within physics and chemistry the use of X-ray diffraction in the determination of crystal and molecular structures has developed steadily from the early work of W. H. Bragg right up to the automated and computerized systems available today. Electron beams at low energies were used by Franck and Hertz in 1914 to investigate the quantized energy states of atoms, and the development of gas discharge tubes, and subsequently thermionic vacuum tubes, provided the first devices for electronics in science and industry. Applications of beta particles and gamma radiation in physics research were at first less spectacular, but the role of alpha particles was of major significance.

To investigate an object, whether micro- or macroscopic, there are basically two available methods - to analyse the information spontaneously or systematically emitted by the object, or to interact with the object under controlled conditions (e.g. to bombard it) and observe the effects produced, including disintegration. The study of radioactivity is a good example of the former and the use of alpha particles to bombard atoms is, even today, a very important example of the latter.

We can distinguish three types of processes arising from the

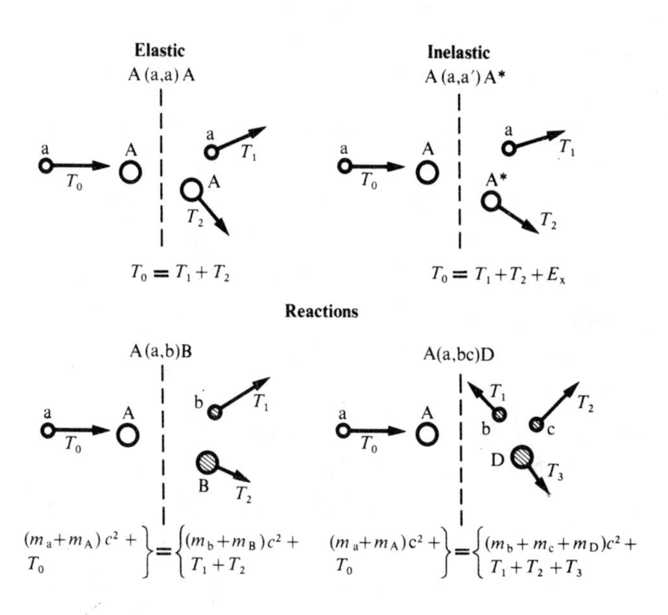

FIG. 3.1. Schematic representation of main types of nuclear interaction processes. The target A is initially at rest. T is the kinetic energy of a particle. For the inelastic scattering, E_x is the excitation energy of A^*. In the energy conservation for the reactions m is the nuclear mass. Linear, and angular momentum are also conserved in all the processes.

bombardment of an isolated object or target A by an incident particle a (Fig. 3.1):

1. Elastic scattering. The two particles retain their identities and their internal energies are unchanged. There is, of course, conservation of energy in the process, and in this case total kinetic energy T is conserved.

2. Inelastic scattering. The two particles retain their identities, but either or both of their internal energies are changed. The change E_x in the sum of their internal energies, by excitation or

de-excitation, is equal to the change in the total kinetic energy.

3. Reactions. The identities of the resultant particles are different from those of the initial particles. One initial particle but not both may retain its identity, with or without a change of internal energy.

Energy conservation must be linked with changes in total mass by Einstein's relation $\Delta E = \Delta mc^2$, as indicated in Fig. 3.1.

ALPHA-PARTICLE SCATTERING

As early as 1906 Rutherford noticed effects due to small-angle scattering of alpha particles in their passage through a thin foil of metal. Later, with Geiger, 'attention was directed to a notable scattering of the α-particles in passing through matter', and the following year (1909) Geiger and Marsden reported more detailed studies. They used thick reflectors (thicker than the range of the alphas) of eight different metals ranging from aluminium to lead, and found that the number of α's scattered at an angle of about 90° increased with atomic weight A. As a result of work with thin foils they comment that 'it seems surprising that some of the α-particles, as the experiment shows, can be turned within a layer of 6×10^{-5} cm of gold through an angle of 90°, and even more. To produce a similar effect by a magnetic field, the enormous field of 10^9 absolute units [10^5 tesla] would be required.' In an estimate of the probability of scatter into backward angles by a thick plate of platinum, they made the simple assumption based on their observations that the scattered alphas 'were distributed uniformly round a half sphere,' and found that 'about 1 in 8000 was reflected under the prescribed conditions.' The restrained comment of Geiger and Marsden ('it seems surprising') was evidently not an adequate description of Rutherford's reaction for in 1936 he declared in his last public lecture, 'it was quite the most incredible event that has ever happened to me in my life. It was almost as incredible as if you had fired a 15-inch shell at a piece of tissue paper and it came back and hit you.' But why was it so surprising?

At the time of the experiments Rutherford accepted J. J. Thomson's theory of the atom as the best working model for estimates of the scattering of both alpha and beta particles. 'The atom is supposed to consist of a number N of negatively charged corpuscles, accompanied by an equal quantity of positive electricity uniformly distributed throughout a sphere.' This quotation is from Rutherford's classic paper on 'The scattering of α and β particles by matter and the structure of the atom' (Philosophical Magazine, 21, 669-88 (1911)), and it should be noted that the apparent agreement between Crowther's measurements of beta scattering and Thomson's theory (1910) was already known to Rutherford. We can reconstruct Rutherford's reasoning using the expressions derived by Thomson.

The average deflection φ_1 due to a single sphere of radius R of positive electricity, is given by

$$\varphi_1 = \left(\frac{1}{4\pi\varepsilon_o}\right)\frac{\pi}{4}\,\frac{NeE}{Mv^2}\,\frac{1}{R}\,,$$

where M, v and E are the mass, velocity, and charge of the incident particle. The average deflection φ_2 due to the N electrons in the atom is given by

$$\varphi_2 = \frac{64}{5\pi}\left(\frac{3}{2N}\right)^{\frac{1}{2}}\varphi_1 = \frac{5.0}{N^{\frac{1}{2}}}\,\varphi_1 \quad .$$

Since the total mean deflection $\varphi = (\varphi_1{}^2 + \varphi_2{}^2)^{\frac{1}{2}} = \varphi_1(1 + 25.0/N)^{\frac{1}{2}}$, the electrons contribute, on Thomson's theory, about 15 per cent of the mean deflection for N = 79 (gold).

The mean-square scattering angle θ_x^2 for a foil of thickness x is φ^2 times the number of atomic collisions. In a foil with n atoms per unit volume, each of area πR^2, the number of encounters is $nx\pi R^2$, hence $\theta_x = (nx\pi R^2)^{\frac{1}{2}}\varphi$. For gold $n = 5.9 \times 10^{28}\ m^{-3}$, and for $x = 2.5 \times 10^{-7}$ m, with $R = 6 \times 10^{-10}$ m, the value of θ_x for 5.5 MeV alpha particles is 3.75°. Geiger (1908) obtained an experimental value of about 3°. However, as Rutherford noted, 'A simple calculation based on the theory of probability shows that the chance of an alpha-particle being deflected through 90° is vanishingly small.' For compound or multiple scattering to produce a deflection at angles greater than θ the probability is $\exp(-\theta^2/\theta_x^2)$, and if θ is set at 90° with $\theta_x = 3^{\circ}$ the probability is about $e^{-900} \simeq 10^{-400}$. The cause of Rutherford's surprise is evident.

It was clear to Rutherford that larger scattering angles would be more probable as a result of single scattering if the radius R of the positively charged sphere were smaller. He therefore considered the limiting case of the charge concentrated at a point. In order to do justice to Rutherford's theory it is convenient to introduce some of the concepts involved and show their relation to measurements and known physical properties.

DIFFERENTIAL CROSS-SECTION

Scattering experiments and their theoretical interpretation are very important in many branches of physics. The probability of scattering is nowadays most commonly expressed in terms of the 'differential cross-section' and the utility and significance of this quantity are best introduced in terms of experimental arrangements and observables, i.e. on an operational basis.

Consider a collimated beam of N_0 particles per second incident on a target of thickness x (Fig. 3.2). A detector

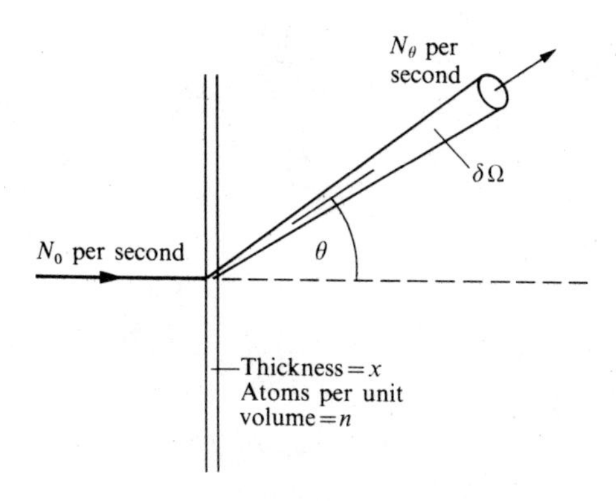

FIG. 3.2. Basic parameters in a nuclear scattering experiment.

subtending a solid angle $\delta\Omega$ at the region of incidence on the target receives N_θ particles per second scattered at a mean angle θ. For thin targets the scattering can be assumed to arise from single deflections, and it is then obvious that N_θ is proportional to N_o, and also to $\delta\Omega$, and also to the number of nuclei in a given area of the target. For reasons discussed later we use the number nx per unit area, where n is the number per unit volume, i.e. N_θ is proportional to N_o nx $\delta\Omega$. The constant of proportionality is the differential cross-section, usually written $d\sigma(\theta)/d\Omega$, hence

$$N_\theta = \frac{d\sigma(\theta)}{d\Omega} N_o \, nx \, \delta\Omega.$$

The interpretation of the words 'differential cross-section' now becomes clear by rearranging the terms

$$\frac{d\sigma(\theta)}{d\Omega} = \frac{N_\theta}{N_o} \frac{1}{nx} \frac{1}{\delta\Omega} \cdot$$

We may define $d\sigma(\theta)/d\Omega$ as the probability N_θ/N_o of scattering at an angle θ into unit solid angle by a target with one atom per unit area. The practical impossibility of a target with one atom per unit area serves to reveal the significance of the phrase 'cross-section', since the probability of the scattering event is the fraction of any given area of the target which is blocked out by the 'effective cross-sectional area' of the nuclei in that area. By choosing unit area of the target for nx and setting $nx = 1$ we obtain a definition of $d\sigma/d\Omega$ that relates directly to the effective cross-section for one nucleus. The units of $d\sigma/d\Omega$ are therefore $m^2\ sr^{-1}$. A common alternative is $b\ sr^{-1}$ where $1b = 1\ \text{barn} = 10^{-28}\ m^2$ and is of the order of the 'geometric' cross-section of the matter distribution in the larger nuclei.

The word 'differential' implies that $d\sigma$ is not the whole of the cross-section. If we confine ourselves to elastic scattering then the total elastic cross-section is

$$\sigma_{e\ell} = \int\left(\frac{d\sigma(\theta)}{d\Omega}\right)d\Omega = \int_0^\pi\left(\frac{d\sigma(\theta)}{d\Omega}\right)2\pi\sin\theta\, d\theta.$$

Thus for a spherically symmetric distribution of scattered particles, as encountered, for example, for low-energy ($10 - 10^4$ eV) neutrons elastically scattered by protons (see pages 125 ff), $\sigma_{e\ell} = 4\pi\, d\sigma(\theta)/d\Omega$. But it is also evident that there are other cross-sections associated with the various possible inelastic scatterings and also the allowable reactions, and these will be large or small according to the probability of the process. Thus ^{238}U has values of σ_{abs} between 10^3 b and 10^4 b for the absorption of neutrons with energies within narrow bands (resonances) between about 6 eV and 100 eV, but the average σ_{abs} in this energy range is about 10 b (Fig. 6.2).

We may summarize the factors that determine the magnitude of $d\sigma(\theta)/d\Omega$.

1. Scattering angle θ - for Rutherford or Coulomb

scattering the function is $\operatorname{cosec}^4(\theta/2)$.

2. Nature of incident particle, notably its mass and electric charge, but also a number of other factors such as its intrinsic angular momentum or spin, and whether these spins are preferentially aligned (i.e. polarized) in relation to the scattering plane.

3. Kinetic energy T_1 of incident particle - for Rutherford scattering the function is $1/T_1^2$.

4. Nature of the target nucleus - similar factors as for the incident particle, including the possibility of aligned spins.

5. Kinetic energy of target nucleus - normally taken as zero except in colliding beam experiments (see pages 76 ff).

6. Type of process: (a) elastic scattering; (b) inelastic scattering (specify excitation energy state(s) involved); (c) reaction (specify in detail).

RUTHERFORD'S THEORY

The only type of process treated at this time by Rutherford was elastic scattering. The simplest possible case is for 180° scattering (Fig. 3.3) and this introduces a useful quantity b_0,

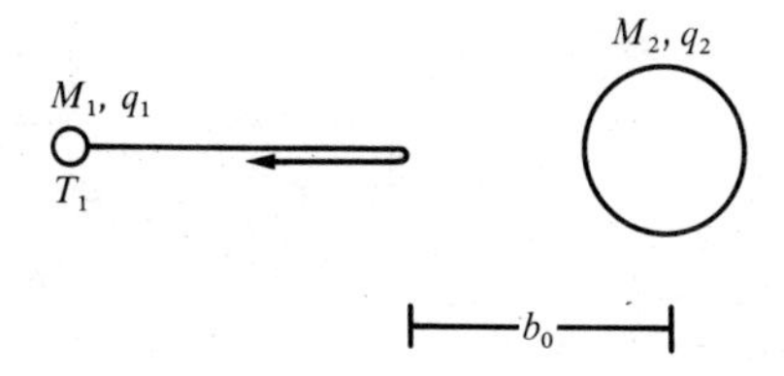

FIG. 3.3. Distance of closest approach, b_0, for 180° scattering.

the distance of closest approach for a head-on collision. At the point of closest approach the kinetic energy T_1 has been changed into electrostatic potential energy

$$T_1 = \frac{1}{4\pi\varepsilon_o} \cdot \frac{q_1 q_2}{b_o}$$

hence

$$b_o = \frac{1}{4\pi\varepsilon_o} \cdot \frac{q_1 q_2}{T_1} .$$

Rutherford obtained an estimate of b_o using $q_2 = 100e$ and $q_1 = 2e$. For gold q_2 is now known to be 79e, and for alphas $q_1 = 2e$. For $T_1 = 5.5$ MeV, $b_o = 41.4 \times 10^{-15}$ m. (10^{-15} m = 1 fm, i.e. one femtometer, but is commonly termed a fermi, after the famous Italian-American nuclear physicist.) Since the radius of the atom is about 10^{-10} m it is clear that the alphas penetrate so deep into the electron distribution that the only effective electric field acting on them is that of the nucleus, at least for scattering angles greater than about 1°.

Rutherford derived an expression for the differential cross-section for elastic scattering at an angle θ on the basis of the following assumptions: (1) the only forces acting are electrostatic; (2) the incident particle does not penetrate the positive charge distribution of the target atom; (3) the electrons in the atom have negligible effect; (4) the recoil of the heavy target nucleus is negligible; (5) the target is thin enough for one target nucleus not to obscure another and for multiple scattering and energy losses to be neglected. The derivation is given in Appendix A, and the result is easy to remember:

$$\frac{d\sigma(\theta)}{d\Omega} = \frac{b_o^2}{16} \operatorname{cosec}^4(\theta/2),$$

where b_o is the distance of closest approach for $\theta = 180^\circ$. For a detector of area D at a distance r from the target,

$$N_\theta = \frac{N_o}{16} \cdot \frac{D}{r^2} \cdot \frac{N_A \rho x}{A} \cdot \left(\frac{q_1 q_2}{T_1}\right)^2 \frac{\operatorname{cosec}^4(\theta/2)}{(4\pi\varepsilon_o)^2},$$

since $n = N_A\rho/A$, where N_A is the Avogadro constant and ρ is the density of the target of atomic weight A.

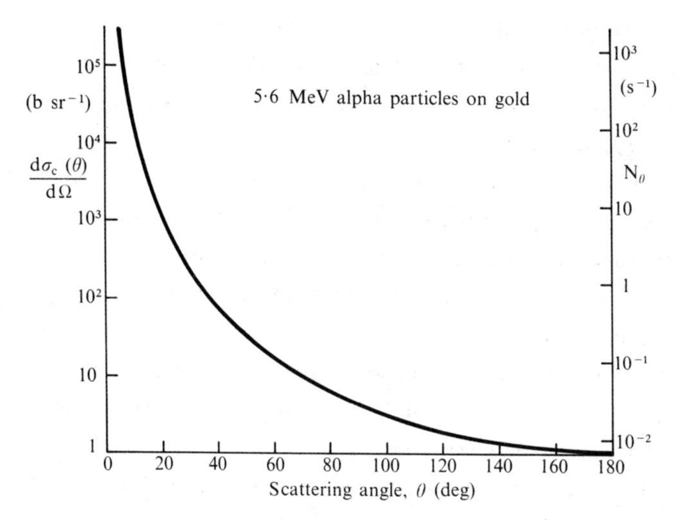

FIG. 3.4. The differential cross-section for Rutherford scattering of 5.6 MeV alphas by gold, and the corresponding count rates for 10^6 alphas per second incident on a gold foil of thickness 10^{-6} m with a detector of area 1 cm^2 placed 30 cm from the target. Notice the logarithmic scale.

In order to impress on the reader the orders of magnitude involved and their dependence on θ, the magnitude of $d\sigma(\theta)/d\Omega$ is shown in Fig. 3.4 for $q_1 = 2e$ (alpha), $q_2 = 79e$ (gold),

$T_1 = 5.6$ MeV. The corresponding N_θ is also shown for the further conditions $N_o = 10^6\ s^{-1}$, $D = 1\ cm^2$, $r = 0.3$ m, ρ (gold) = 19 300 kg m^{-3}, A (gold) = 197, x = 10 μm. Note that the vertical scales are logarithmic.

Rutherford reported in the same paper that Geiger had found that measurements between 30° and 150° for gold were in agreement with the theoretical angular distribution. Later Geiger and Marsden (1913) confirmed this between 5° and 150° and also showed that N_θ was proportional to the thickness x and to the inverse square of the kinetic energy T_1. Over 100 000 scintillations were individually counted! (See pages 88 ff). The relation to q_2 was more difficult to determine as the absolute numbers of scattered particles had to be measured. Nevertheless the nuclear charge number q_2/e for gold was estimated to be within 20 per cent of half the atomic weight; the accepted values are $q_2/e = 79$, A = 197.

The conclusions to be drawn from this work are as follows.

1. The major part of an atom (i.e. all except the relatively light electrons) is concentrated in a nucleus with a positive electric charge equal to approximately (A/2)e.

2. The interaction between incident alpha particles of energies between about 2 MeV and 6 MeV and nuclei ranging from aluminium to gold can be adequately explained by the electrostatic force alone, and the alpha particle at its point of closest approach has not penetrated the charge distribution of the target nucleus. We may interpret this as setting an upper limit for the radius of the nucleus $R < b_o = 41.4 \times 10^{-15}$ m for gold, and $R < b_o = 6.8 \times 10^{-15}$ m for aluminium.

Rutherford's theoretical interpretation of Geiger and Marsden's experiments firmly established the picture of the nuclear atom, thereby paving the way for theories of atomic structure and detailed studies of the nucleus itself.

DIFFRACTION EFFECTS IN ALPHA SCATTERING

The next major development in the study of the elastic scattering of heavy charged particles (i.e. alpha particles and, later, protons, etc.) came with the construction of nuclear accelerators, in particular the simple cyclotron (see pages 64 ff). Incident particles of higher energies approach more closely to the nucleus, and at a certain distance the Rutherford formula begins to be inaccurate. This occurs at a particular energy for observations at a fixed angle (Fig. 3.5(a)), and correspondingly for an adequately high energy becomes evident at a particular angle (Fig. 3.5(b)). The effects are more noticeable for lighter

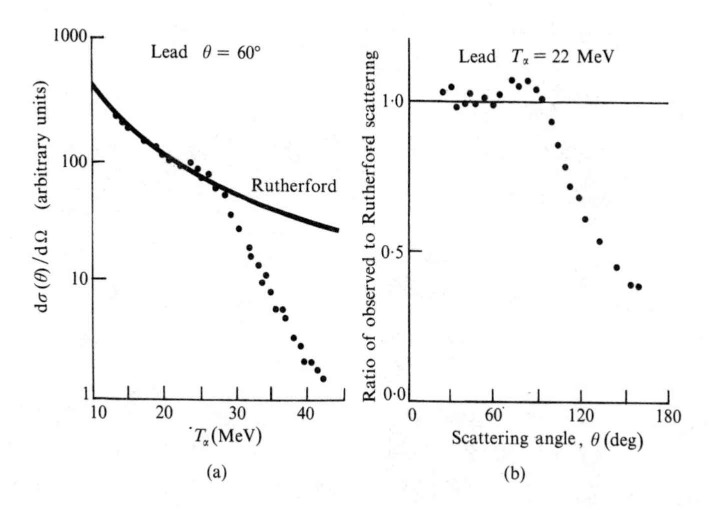

FIG. 3.5. Limits of application of Rutherford's scattering formula. (a) (Adapted from G. W. Farwell and H. E. Wegner, Phys. Rev. 95, 1212 (1954).) (b) (Adapted from N. S. Wall, J. R. Rees and K. W. Ford, Phys. Rev. 97, 726 (1955).)

atoms with their lower nuclear charge Ze. At a high enough energy an important new phenomenon is found for nearly all incident particles and target nuclei, namely, an angular distribution characteristic of diffraction effects (Fig. 3.6).

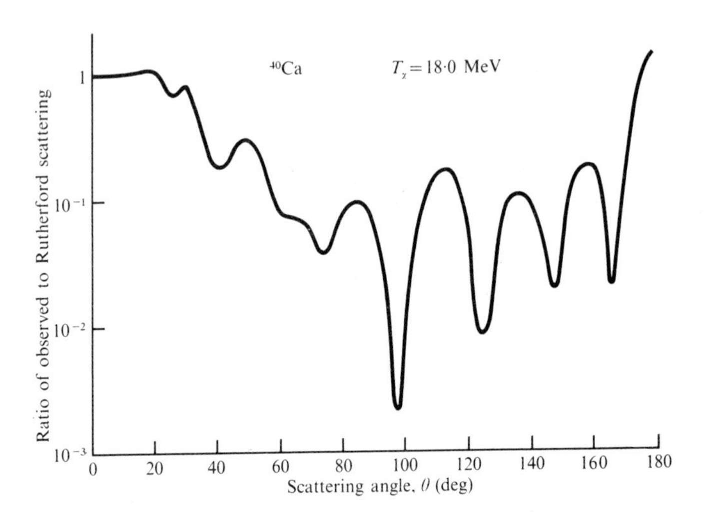

FIG. 3.6. Diffraction pattern observed for 18 MeV alpha particles scattered by ^{40}Ca. (Adapted from C. P. Robinson, J. P. Aldridge, J. John and R. H. Davis, Phys. Rev. 171, 1241 (1968).)

The interpretation in terms of the wave nature of moving particles, using the de Broglie wave-length $\lambda = h/Mv$ (see Radiation and quantum physics (OPS 3) by D. J. E. Ingram), yields further information on the size of the nucleus. Thus for the simplest possible model, in which the nucleus is treated as a black disc of radius R the angle of the first diffraction minimum is $\theta_{min} \simeq \sin^{-1}(0.61\lambda/R)$. Hence for $\lambda = R$, $\theta_{min} \simeq 38^{\circ}$. Since $\lambda_\alpha = h(2M_\alpha T_\alpha)^{-\frac{1}{2}}$, $\lambda_\alpha = 4.0$ fm at $T_\alpha = 18$ MeV, suggesting $R \simeq 5.5$ fm for ^{40}Ca (Fig. 3.6) for which $\theta_{min} \simeq 26^{\circ}$. A more sophisticated treatment is provided by the optical model of the nucleus (see The properties of nuclei (OPS 12) by G. A. Jones) which allows for reflection, refraction, and absorption ('cloudy crystal ball') and introduces a surface diffuseness in place of a hard, abrupt surface. Its ability to account for both elastic and inelastic scattering, and for polarization effects, is well established for both charged and uncharged particles. The required size and shape of the nuclear matter distribution is closely related to the electric charge distribution deduced from comparable experiments with high-energy electrons (T_e is a few hundred MeV) and the commonest simple function used to describe it is the so-called Fermi or Saxon-Woods formula,

$$\rho_r = \frac{\rho_o}{1 + \exp\{(r - R)/a\}},$$

where ρ is the matter or charge density. This is illustrated in Fig. 3.7 for calcium (A = 40) and gold (A = 197) using values for R obtained from the important relation deduced from a wide range

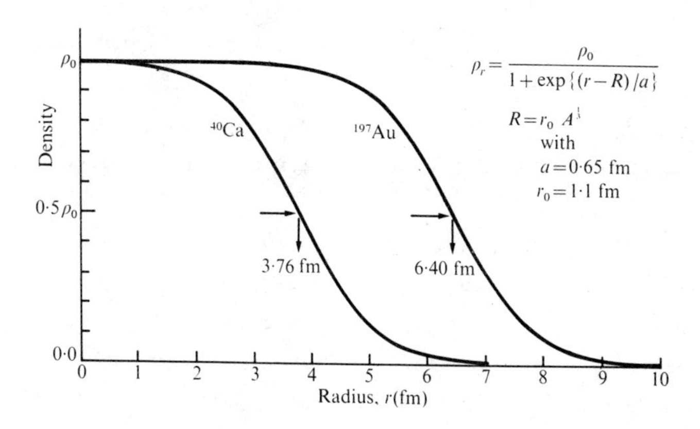

FIG. 3.7. Radial dependence of nuclear charge density given by Saxon-Woods (or Fermi) formula with parameters r_o and a obtained from charged particle scattering

and variety of experiments,

$$R = r_o A^{\frac{1}{3}} ,$$

with $r_o = 1.1$ fm. The diffuseness parameter a is set at 0.65 fm and it can be shown that the surface region from 10 per cent to 90 per cent maximum density extends over $4.4a \simeq 3$ fm. The values of r_o and a for n, p, d, ^3H, ^3He, and alpha scattering, for energies up to at least 100 MeV, do not need to be varied by more than about 15 per cent from the values used here. It is clear from the Fermi formula that R is the 'half-density' radius, and to a good approximation the central charge and matter densities and also the surface thickness are the same for all nuclei, excepting the very lightest.

ALPHA-PARTICLE SPECTRA

The first indication that the alpha particles emitted by a given species of nucleus were not all of the same energy was noted by Rutherford and Wood (1916). They found that about 1 in 10^4 alphas from a thorium C (^{212}Bi) source had a range of some 113 mm in air, (equivalent to $T_\alpha = 10.6$ MeV) and about 3 in 10^5 had a 100 mm range. The principal group had a range of 86.2 mm (8.8 MeV). Comparable long-range alphas were also found for radium C (^{214}Bi).

As late as 1930, Rutherford, Chadwick, and Ellis, in their classic volume 'Radiations from radioactive substances', were unable to decide between two possible origins: (1) a 'separate radioactive product' with an exceedingly short decay period, as deduced from the Geiger-Nuttall relation, and (2) an extra supply of energy given to the normal alphas 'just before their ejection'. They were percipient enough to note that the extra energy was approximately equal to the energies of gamma rays from radioactive nuclei.

Rosenblum (1929) used a powerful electromagnet to deflect a beam of alpha particles through 180°, thereby achieving a focusing effect that facilitated accurate measurements of the radii of curvature of the tracks of the alphas, and hence their energies. He found not only the rare high-energy alphas but also groups of lower energies. In particular for thorium C (^{212}Bi) six groups are found at 6.086 MeV (27.2%), 6.046 MeV (69.9%), 5.758 MeV (1.7%), 5.615 MeV (0.15%), 5.603 MeV (1.0%), and 5.478 MeV (0.014%).

By a close analogy with the spectral lines from excited atoms, Ellis (1922) explained various relations between gamma-ray energies (e.g. $h\nu_1 + h\nu_2 = h\nu_3$) in terms of excited states of nuclei. It was later realized that the differences between the alpha-particle energies for thorium C corresponded to the observed energies of associated gamma rays (e.g. 40 keV, 328 keV, 471 keV), and the careful construction of energy-level schemes finally led to the representation shown in Fig. 3.8. On this diagram the energies of the emitted alpha particles are indicated, but in order to calculate accurately the corresponding energy levels and the consequent gamma rays it is essential to allow for the recoil of the daughter nucleus. Thus for the 6.086 MeV alpha particle the ^{208}Tl recoil energy is 117 keV, and for the 5.603 MeV alpha it is 108 keV.

The explanation of the long-range alphas now follows naturally - the nucleus (^{212}Po) is excited (e.g. by 1.797 MeV) when the alpha is emitted, and the total available energy is therefore that much greater. Most excited states of nuclei decay by gamma emission in times of the order of 10^{-12} s. For ^{212}Po in the ground state the half-life for alpha emission is 3×10^{-7} s and for the excited state the half-life is much shorter. There is thus competition between gamma emission and alpha emission. This explanation sits astride the two possibilities of Rutherford, Chadwick, and Ellis if we allow that ^{212}Po excited by 1.797 MeV

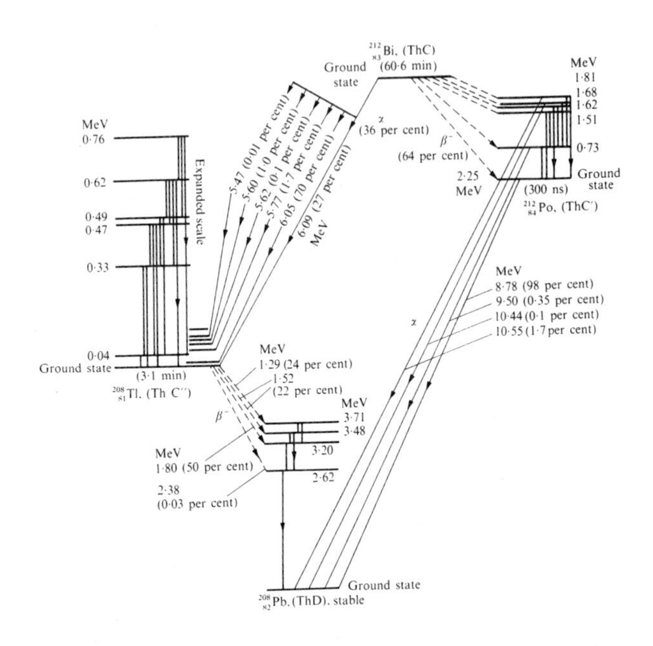

FIG. 3.8. Decay Scheme for ^{212}Bi with alpha and beta energies, and the subsequent gamma energies, correlated in the energy levels of the daughter and parent nuclei.

is a 'separate radioactive product'.

A number of features are illustrated in Fig. 3.8.

1. Branching of a radioactive series - ^{212}Bi decays 36% by alpha emission and 64% by beta emission.

2. Emission of betas of several maximum energies by the same nuclear species, with correlated gamma rays from the excited daughter.

3. The stable end product ^{208}Pb. This is a nucleus with a number of remarkable properties. Note, for example, the very large value (2.62 MeV) of its

first excited state compared with 0.040 MeV (^{208}Tl) and 0.726 MeV (^{212}Po). (^{208}Pb is 'doubly magic', Z = 82, N = 126 (see pages 134 ff).)

4. In balancing energy changes for the alternative routes between ^{212}Bi and ^{208}Pb it is essential to use maximum beta energies.

5. 'Forbidden' gamma transitions, e.g. the ^{208}Tl level at 0.492 MeV cannot decay directly to the ground state. The gammas in time-coincidence with the 5.62 MeV alpha have energies 0.452 MeV and 0.040 MeV.

ALPHA EMISSION

Geiger and Nuttall (1911) proposed a relation (see page 24) between the energy of emission of alpha particles T_α and the radioactive constant λ of the parent nucleus, $\ln\lambda = A_1 \ln T_\alpha + B_1$, where A_1 and B_1 are constants. The derivation of a closely similar relation, in 1928, was one of the early triumphs of wave mechanics, and involved the recognition that alpha particles represented as a wave have a finite probability of being transmitted through the potential barrier at the surface of the nucleus, even when the particle energy is less than the height of the barrier.

This potential barrier is produced by the combination of the nuclear (attractive) potential and the Coulomb (repulsive) potential. A realistic form for $V(r) > R$, as obtained from alpha-scattering experiments (see page 50) by Igo (1958), is shown in Fig. 3.9. Also shown is a simplified form with the usual Coulomb potential $V_C = zZe^2/4\pi\varepsilon_0 r$ cut off at the nuclear radius $R = r_0 A^{\frac{1}{3}}$ by a square-well nuclear potential. Neither of these forms is simple enough to encourage an analytic solution of the wave equation

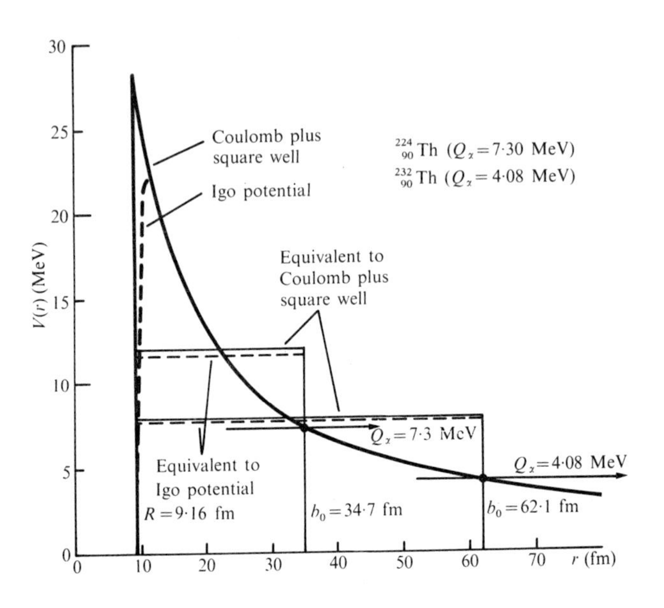

FIG. 3.9. Potential barriers for thorium used in the analysis of alpha emission probabilities. The equivalent square barriers are calculated to give the same transmission probabilities as those of the realistic Igo potential and the 'Coulomb plus square-well' potential, and use barrier widths of b_o-R for the Q_α = 7.3 MeV and 4.08 MeV, where b_o is the distance of closest approach for 180° scattering by the daughter nucleus.

$$d^2\psi/dr^2 + [2\mu_\alpha\{Q_\alpha - V(r)\}/\hbar^2]\psi = 0$$

where $\mu_\alpha = M_\alpha(A-4)/A$ is the reduced mass of the alpha particle emitted by a nucleus of mass number A, and Q_α is the total available energy of the alpha and the nuclear recoil, $Q_\alpha = T_\alpha A/(A-4)$. We shall therefore use a square barrier, shown in Fig. 3.10(a), and first treat the problem in one dimension. Notice the simplifying assumption that V(x) = 0 for $x < -b$, corresponding to inside the nucleus, where the form of the

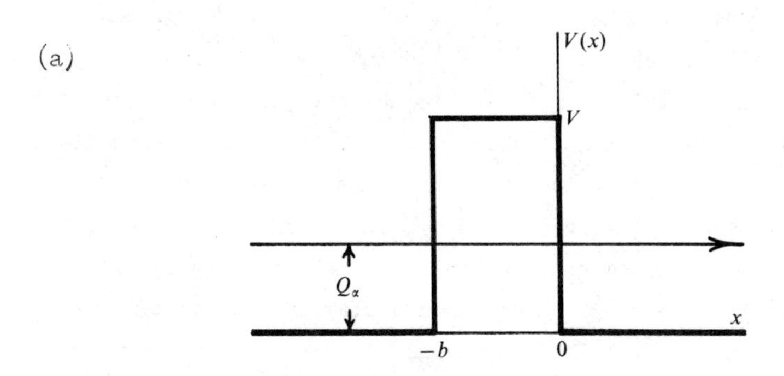

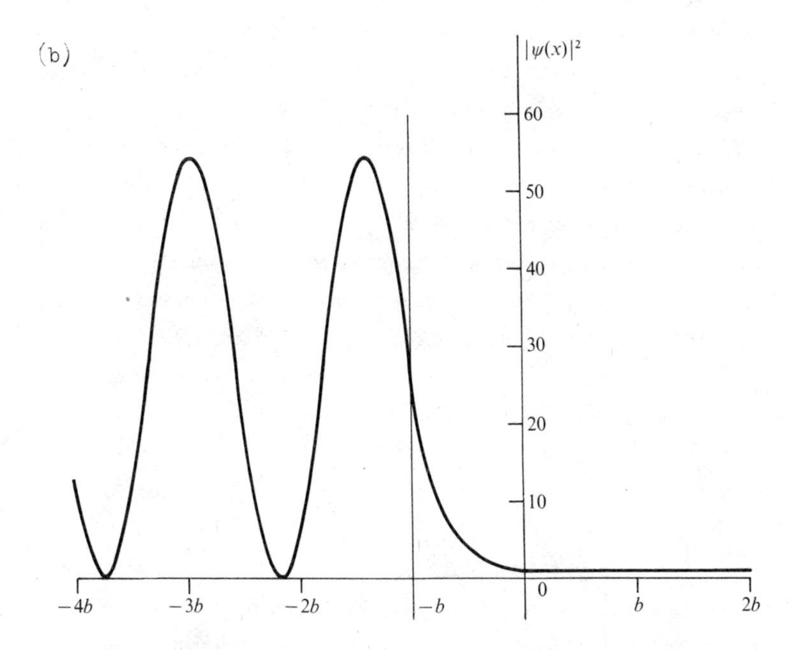

FIG. 3.10. (a) Square barrier used in simple theory of alpha emission. (b) corresponding probability per unit length, $|\psi(x)|^2$, in the three regions, calculated for $Q_\alpha = V/2$, unit transmitted intensity, and $b = 2/\alpha$ (see text).

potential is not at all well established. The solution, derived in Appendix B, is shown in Fig. 3.10(b) for $Q_\alpha = V/2$ and $b = 2/\alpha$, where the probability $|\psi|^2$ of finding the alpha particles in unit length is plotted in relation to the barrier. An interpretation of Fig. 3.10(b) is not difficult. The oncoming wave, exp $(i\alpha x)$ with $\alpha^2 = 2\mu_\alpha Q_\alpha/\hbar^2$, is partially reflected and partially transmitted at the front of the barrier $(x = -b)$. The 'wave' that enters the barrier is partially reflected at the far side of the barrier $(x = 0)$ and hence multiple reflections take place at $x = -b$ and $x = 0$. The sum of the reflected waves for $x < -b$ is $A \exp(-i\alpha x)$ and combines with the incident wave to produce a standing wave and therefore a sinusoidally varying intensity superimposed on a finite flux. The sum of the 'waves' within the barrier (of the forms $\exp(-\beta x)$ and $\exp(\beta x)$ with $\beta^2 = 2\mu_\alpha (V - Q_\alpha)/\hbar^2$) yields a probability of penetrating the barrier which decreases steadily (not exponentially) with the depth of penetration until the width of the barrier is traversed. The sum of the transmitted waves ($D \exp(i\alpha x)$) corresponds to a steady beam, i.e. a constant probability as a function of x. The flux throughout is clearly independent of x, otherwise there would be a local pile-up or absorption of particles.

The probability of transmission through the barrier (see Appendix B) is given by

$$\text{transmission} = 4(Q_\alpha/V)(1 - Q_\alpha/V)(1 + \sinh^2\beta b)^{-1}.$$

In order to apply this result to alpha emission we need to make further simplifying assumptions. We first note that the three-dimensional solution $\psi(r)$ for a spherically symmetric nucleus, emitting an alpha along a radius, can be written $(1/r)U(r)$, where $U(r)$ turns out to be identical in form with $\psi(x)$, and we simply substitute r for x and U for ψ. A glance at Fig. 3.9 suggests that suitable values for V and b of the equivalent square-barrier potential are about 15 MeV and 30 fm respectively for $Q_\alpha = 7.30$ MeV, and for these values $\beta b = 36$. It follows that for such large values of βb a very good approximation is $(1 + \sinh^2\beta b)^{-1} = 4 \exp(-2\beta b)$. We now make the very crude and crucial assumptions, suggested by Fig. 3.9, that $V \simeq 2Q_\alpha$ for $b \simeq b_o$, where b_o is the distance of closest approach. Since $b_o = zZe^2/4\pi\varepsilon_o Q_\alpha$, with $z = 2$ and Z for the residual nucleus, and $\beta = (2\mu_\alpha/\hbar^2)^{\frac{1}{2}}(V-Q_\alpha)^{\frac{1}{2}}$, it follows that $2\beta b = KQ_\alpha^{-\frac{1}{2}}$ for $K = (8\mu_\alpha)^{\frac{1}{2}} zZe^2/4\pi\varepsilon_o\hbar$. For this simple square-barrier theory the probability of transmission is therefore $4 \exp(-KQ_\alpha^{-\frac{1}{2}})$.

The radioactive constant λ is the probability of transmission multiplied by the frequency of impact on the barrier. This frequency can be set approximately equal to the velocity of alphas within the nucleus, say 10^7 m s^{-1}, divided by the nuclear radius, say 10^{-14} m, i.e. alphas make about 10^{21} attempts per second to escape, and this value is roughly independent of Q_α and A. Hence λ is proportional to $\exp(-KQ_\alpha^{-\frac{1}{2}})$, or $\ln\lambda = -KQ_\alpha^{-\frac{1}{2}} + B_2$, where B_2 is approximately constant. Detailed treatments with realistic shapes for the potential barrier yield closely similar relations, and plots of measured values of $\ln\lambda$ against $Q_\alpha^{-\frac{1}{2}}$ (or $T_\alpha^{-\frac{1}{2}}$) are very nearly straight lines. This theoretical relation is different from that proposed by Geiger and Nuttall, $\ln\lambda = A_1 \ln T_\alpha + B_1$, but it happens that, for the ranges of T_α involved, the form of $T_\alpha^{-\frac{1}{2}}$ is quite close to that for $\ln T_\alpha$, as shown in Fig. 3.11. This figure also includes the form of the exact solution of the Coulomb potential with the cut-off at radius R (Fig. 3.9), for which the relation is

$$\ln\lambda = -KQ_\alpha^{-\frac{1}{2}}\{\cos^{-1}(R/b_o)^{\frac{1}{2}} - (R/b_o - R^2/b_o^2)^{\frac{1}{2}}\} + B_3.$$

Notice that this formula depends on R and allows estimates of this parameter which yield $R \simeq 1.5\, A^{\frac{1}{3}}$ fm, some 25 per cent larger than presently accepted values. A common, but severe,

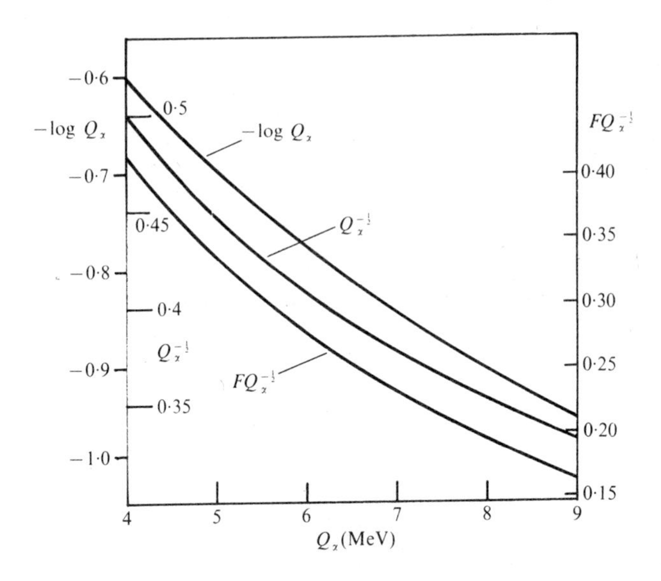

FIG. 11. The three functions encountered in versions of the Geiger-Nuttall relation, showing their close similarity for Q_α = 4 -9 MeV. Log Q_α is from the original empirical relation, $Q_\alpha^{-\frac{1}{2}}$ from the simple square-barrier theory and $FQ_\alpha^{-\frac{1}{2}}$ from the Coulomb potential with square cut-off. (The Geiger-Nuttall relation includes an adjustable constant that allows the arbitrary vertical displacement used to prevent confusing overlaps of the curves.)

approximation is to consider $R/b_0 \ll 1$, whence the term in square brackets reduces to $\pi/2$, with a return to the form

$\ln \lambda = A_2 Q_\alpha^{-\frac{1}{2}} + B_2$.

The interpretation of the sensitive dependence of the half-life $\tau_{\frac{1}{2}}$ (= 0.693/λ) on the alpha energy T_α follows immediately. Experiments show that $^{232}_{90}$Th has $\tau_{\frac{1}{2}} = 4.45 \times 10^{17}$ s for T_α = 4.01 MeV (Q_α = 4.08 MeV) and $^{224}_{90}$Th has $\tau_{\frac{1}{2}}$ = 1 s for T_α = 7.17 MeV (Q_α = 7.30 MeV). In the square-barrier theory with $b = b_0$ we note that the multiplying factor in the transmission $16(Q_\alpha/V)(1-Q_\alpha/V)$ is unimportant compared with the exponential term. It has a maximum value of 4 for $Q_\alpha = V/2$ and is 3 for $Q_\alpha/V = 1/4$ and $3/4$. Hence $\tau_1/\tau_2 \simeq \exp(2\beta_1 b_1 - 2\beta_2 b_2) = \exp\{K(Q_1^{-\frac{1}{2}} - Q_2^{-\frac{1}{2}})\}$. But $K \simeq 219$ for Q measured in MeV hence $\tau_1/\tau_2 \simeq 10^{12}$ which, owing to the many approximations, is not in good agreement with experiment, although it approaches the correct order of magnitude. For the truncated Coulomb potential the value is $\tau_1/\tau_2 \simeq 8.8 \times 10^{17}$, and it is close to 5×10^{17} for the Igo potential (Fig. 3.9), in excellent agreement with observations. The absolute values of τ_1 and τ_2 require more careful consideration of the approximations than does the ratio τ_1/τ_2 and the simple square-barrier approach with $b = b_0$ is again inadequate ($\tau_1 \simeq 10^{26}$ s and $\tau_2 \simeq 10^{14}$ s). But for the truncated Coulomb potential the half-lives are in rather good agreement with experiment ($\tau_1 = 4.1 \times 10^{17}$ s and $\tau_2 = 0.45$ s), assuming that the frequency of impact on the barrier is 10^{21} s^{-1}. In Fig. 3.9 the equivalent square-barrier potentials producing the same transmissions are shown for $b = b_0 - R$. The overall success of this wave-mechanical interpretation of alpha emission is most impressive (see G. A. Jones, in this Series).

The probability of alpha emission is directly related to the probability for the inverse process, alpha absorption. The same theory also applies to the emission and absorption of other charged particles, e.g. protons or ${}^{7}_{3}Li$ nuclei, with suitable changes of parameters. For example, if it were energetically allowed the emission of ${}^{7}_{3}Li$ with the same energy (4 MeV) and the same residual nucleus, which will be ${}^{228}_{88}Ra$, would have a half-life of about 4×10^{60} years, indicating an unobservably low level of activity. (This value is obtained for the truncated Coulomb potential.) The similar calculation for the emission of a 4 MeV proton, leading to the same residual nucleus, yields a half-life of 1.6×10^{-15} s. This is of the same order of magnitude as the lifetime of the compound nucleus formed when an incident particle coalesces with a target nucleus. Notice that so-called 'proton radioactivity' occurs when beta decay leads to a proton-emitting nucleus, and can have (beta decay) lifetimes much larger than 10^{-15} s. By analogy to 'delayed neutron emission' (see page 109) it is better to call this '(beta) delayed proton emission'.

ALPHA-PARTICLE REACTIONS

One of the problems solved by the wave-mechanical interpretation of alpha emission (1928) was how it came about that 5 MeV alphas (say) were observed to be emitted by a given nucleus but even 7 MeV alphas were not observed to be absorbed by that nucleus or by nuclei of comparable A and Z. It was noted above that for heavy nuclei the probability of exit (or entry) of a 7 MeV alpha is about 1 in 10^{21}, but experiments to measure such small absorptions are notoriously difficult to undertake.

However, in 1917, long before wave mechanics, the effects of alpha-particle absorption reactions had been observed and correctly interpreted by Rutherford. Marsden noted in 1914 in the course of scattering experiments that alpha particles incident on hydrogen gas produced long-range particles. The particles passed through a foil that was thick enough to stop the alphas and then produced scintillations on a zinc-sulphide screen. They were considered to be protons, i.e. knock-on hydrogen nuclei from elastic collisions. Rutherford repeated the experiments with various gases and reported 'On introducing oxygen or carbon dioxide into the vessel, the number of scintillations fell off in amount corresponding with the stopping power of the column of gas. An unexpected effect was, however, noticed on introducing dried air ... the number of scintillations increased ... which appeared of about the same brightness as [hydrogen]-scintillations.' After further investigations he boldly surmised 'it is difficult to avoid the conclusion ... that the nitrogen atom is disintegrated under the intense forces developed in a close collision with an alpha particle.' The closeness is indicated by the formula for the distance of closest approach, $b_o = zZe^2/4\pi\varepsilon_o T_\alpha$, where Z is small for the light elements (b_o = 3.5 fm for T_α = 5.6 MeV on nitrogen). All the elements from boron to potassium, apart from carbon, oxygen, and, possibly, beryllium, emitted protons under alpha bombardment.

Proposed interpretations included (1) the alpha knocks out a proton from the nucleus and itself remains an alpha; (2) the alpha is absorbed and a proton is emitted; (3) the alpha is broken up and the nucleus unchanged; and (4) both alpha and nucleus are broken up. In 1925, Blackett used a cloud chamber (see pages 91 ff) to make visible the tracks of the charged particles involved including the recoil nucleus, and showed unambiguously from 8 examples out of 400 000 alpha collisions that (2) is the most probable. We may write the nuclear reaction equation

$$^{14}_{7}N + {}^{4}_{2}He \rightarrow \left({}^{18}_{9}F^*\right) \rightarrow {}^{1}_{1}H + {}^{17}_{8}O,$$

taking care to balance the electric charges of the nuclei involved and select the obvious mass numbers. The intermediate step, in which an excited 'compound nucleus' is formed (* indicates excitation), has been found to take place in many such reactions. The overall reaction is sometimes summarized as $^{14}N\ (\alpha, p)^{17}O$, and from the conservation of mass-energy we can deduce the energy of the proton for a given alpha-particle energy (see pages 100 ff) if we know sufficiently accurately the masses of the nuclei involved.

The significance of Rutherford's observations was not under-estimated. This artificial transmutation of the elements had been claimed without adequate foundation not only by alchemists but also by nineteenth-century scientists. The wave-mechanical interpretation of alpha emission indicated how very much more probable would be the entry of a proton into the nucleus, and this stimulated the building of (proton) accelerators. The first proton reaction was observed with ^{7}Li in 1932, by Cockcroft and Walton using 125 keV protons.

The important role of alpha-particle reactions was not yet eclipsed, for in the same year Chadwick introduced the neutron to explain some highly penetrating radiation observed by Bothe and Becker (1932) when beryllium was exposed to alphas from polonium. The neutron was discussed as a possibility by Rutherford and others as early as 1920 and even systematically sought by Chadwick before 1930. In 1932 Irene Joliot Curie and Frederick Joliot discovered that Bothe and Becker's 'radiation' ejected protons of high velocity from a sheet of hydrogenous material, e.g. paraffin wax. The conservation laws of energy and momentum excluded the possibility that the radiation consisted of very high-energy gamma photons, and Chadwick proposed 'particles of mass nearly equal to that of the proton and with no net charge' The neutron's lack of electric charge prevents it from producing ionization along its track, and explains why it was relatively difficult to detect and identify. It also suggests that it should react readily with nuclei since there would be no Coulomb potential barrier, and this was found to be the case (see pages 105 ff).

Alpha-particle scattering led to the discovery of the nucleus, and alpha-particle reactions led to the discovery of the neutron, neutron reactions, and nuclear power generators. The discoveries of Rutherford's generation not only became the tools of the next generation but also provided one of its sources of power.

PROBLEMS

3.1. Check Geiger and Marsden's calculation (see page 38) of the magnetic flux density required to turn alpha particles through $\theta > 90^\circ$ in a layer of thickness $d = 6 \times 10^{-7}$ m. List the assumptions and approximations made.

3.2. Calculate $d\sigma(\theta)/d\Omega$ for Coulomb scattering at $\theta = 90^\circ$ for 5.6 MeV alpha particles on gold, and hence the number scattered from an incident beam of 10^6 alphas per second into a detector of area 1 cm^2 placed 0.3 m from the target of thickness 10^{-6} m (Fig. 3.4).

3.3. A small source of alpha particles, activity S, and a small semiconductor detector, area D, lie symmetrically on the axis of, and on opposite sides of, a narrow annular transmission target, radius R, width δR, thickness x, n nuclei per unit volume. Show that the count-rate for scattering at an angle θ is $(S\delta RnxDb_o^{\,2}/32R^3)\cos(\theta/2)$, where b_o is the distance of closest approach of the alphas to the nucleus for 180° scattering.

3.4. In an experiment on Coulomb scattering, replacement of a thin ${}^{64}_{30}Zn$ target by a ${}^{114}Cd$ target of the same mass per unit area increased the scattered intensity by a factor 1.45. Calculate the atomic number of cadmium.

3.5. Calculate the recoil energy of (a) the ${}^{208}Tl$ nucleus when a 6.086 MeV alpha particle is emitted from ${}^{212}Bi$, and (b) the ${}^{137}Ba$ nucleus when a 660 keV gamma photon is emitted. Use integral masses for (a) and M(Ba) = 137 u for (b).

4. Nuclear Accelerators

INTRODUCTION

A great deal was discovered about the nucleus by the use of naturally occurring alpha particles, but their maximum energy is only about 5.5 MeV for useful source strengths, and their double electric charge inhibits their near approach to all but the lightest nuclei. Machines were therefore designed to accelerate stable charged particles of any type, and the highest energies obtainable have become limited only by the cost associated with the consequent large size and complexity.

In practice, the experimental nuclear physicist is concerned with the characteristics of the beam of charged particles. These include (1) type(s) of particle, (2) energy and energy resolution, (3) spatial and angular distribution, (4) intensity, (5) time structure (see below), (6) polarization, and (7) stability of the factors (1) to (6). These characteristics are determined by the initial source of ions, the type of accelerator, and the beam transport system. The ion source usually involves a radio-frequency discharge and determines the type of particle, the intensity, and the polarization, i.e. the degree of orientation of the spins of the particles. The type of accelerator determines the energy and energy spread, the spatial and angular distribution, and the time structure (e.g. continuous current, pulses of particles, or pulses with a fine radio-frequency bunching). Beam transport is an important part of the process, especially in experiments with beams of unstable fundamental particles, (e.g. pions, kaons) which have to be produced by bombarding targets, sometimes within the accelerator, with very high-energy stable particle (e.g. protons). Selection of the desired type of particle is then achieved by magnetic and

TABLE 4.1.

	Particles	Maximum energy or voltage
1. Continuous acceleration		
(a) Cockcroft-Walton (voltage multiplier)	Any	≃ 3 MV
(b) van de Graaff - single (electrostatic)	Any	≃ 12 MV
van de Graaff - tandem (two stages with electron stripping half way)	Start with negative ions	≃ 24 MV
(c) Betatron (magnetic induction)	e	≃ 400 MeV
(d) Electron LINAC (travelling electromagnetic wave in waveguide)	e	(20 GeV)
2. Repeated acceleration using radio-frequency fields		
(a) Linear accelerator (resonant cavity with drift tubes)	Any	(800 MeV p's)
(b) Simple cyclotron (magnetic resonance)	Any	≃ 22 MeV p's ≃ 80 MeV α's
(c) Sector-focusing cyclotron (fixed frequency)	Any	(600 MeV p's)
(d) Synchrocyclotron (frequency-modulated)	Any	≃ 700 MeV p's
(e) Synchrotron - proton (vary B and possibly f)	p	(500 GeV)
(f) Synchrotron - electron (vary B)	e	(6 GeV)

(Energies in brackets may be exceeded by increasing size).

electric separators, based on principles that are similar to those of mass spectrometers. Together with magnetic deflectors, collimators, and a series of magnetic lenses, these particle separators also influence the band of energies selected, the transmitted intensity, and (within limits) the spatial and angular distribution.

There are two main types of accelerator. One type produces an accelerating electric field acting continuously on a bunch of particles or on a steady current of particles. The other type produces repeated accelerations in regions localized in space. The repetition implies alternating electric fields, and it is therefore necessary to shield the particles during intervals of unfavourable field directions or phases. A summary of the principal kinds of accelerator is given in Table 4.1.

CONTINUOUS ACCELERATION

The Cockcroft-Walton voltage-multiplier consists of rectifiers and capacitors (Fig. 4.1) and was the first accelerator (1930) used in nuclear-reaction studies. It is still often used as the first stage of even the largest accelerators. Its action is as follows: When B is positive with respect to A, it is clear that C_1 will tend to charge up, through the diode BC, to a maximum voltage V. If D is at a lower potential than that of C, then current will flow through diode CD and will tend to charge up C_2 and also provide current for diode DE, depending on the voltage already on C_2 and on the potential difference across DE. When B is negative with respect to A, the voltage acting across CB is eventually, after many cycles, a maximum of +V from C_1 and +V from the transformer secondary, i.e. C_2 will tend to charge up to 2V through diode CD. Similarly, when B is positive with respect to A, the voltage acting across DC is eventually a maximum of +2V from C_2, -V from the transformer secondary, and +V from C_1, i.e. C_3 will tend to charge up to 2V through diode DE.

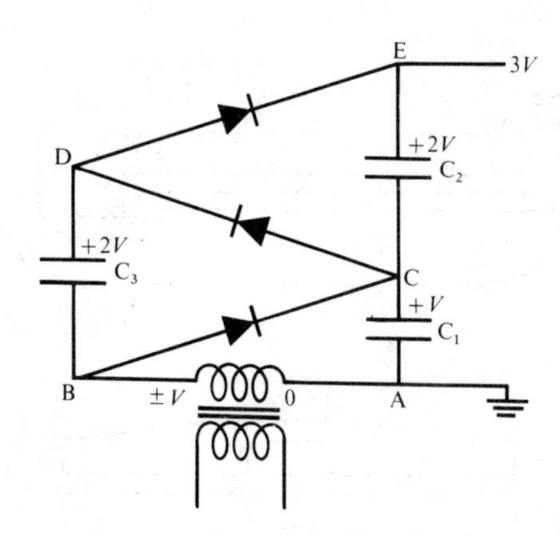

FIG. 4.1. Circuit diagram to illustrate principle of the Cockcroft-Walton voltage multiplier.

Allowance must be made for current drain from E to A, associated with the accelerated ions and inevitable insulation losses, and the maximum voltage is, therefore, a little less than V times the number of rectifiers.

The simple van de Graaff accelerator (Fig. 4.2(a)) consists basically of a moving loop of insulating material, at one end of which charge is sprayed on by a 'comb' of points maintained at some 30 kV. At the other end, inside a large hollow rounded conducting terminal, the charge is removed by a similar device. The removal process is an example of the Faraday-cage effect whereby a charged conductor inside a larger hollow conductor (the 'cage') will, when joined to the hollow conductor, transfer its charge whatever the potential of the cage. The positive ion source is located in this high-voltage terminal, and the ions are accelerated in a suitable evacuated tube down to earth potential. In order to minimize the main voltage limitation, namely corona discharges to laboratory walls and nearby objects, the whole system is commonly enclosed in a pressurized steel tank

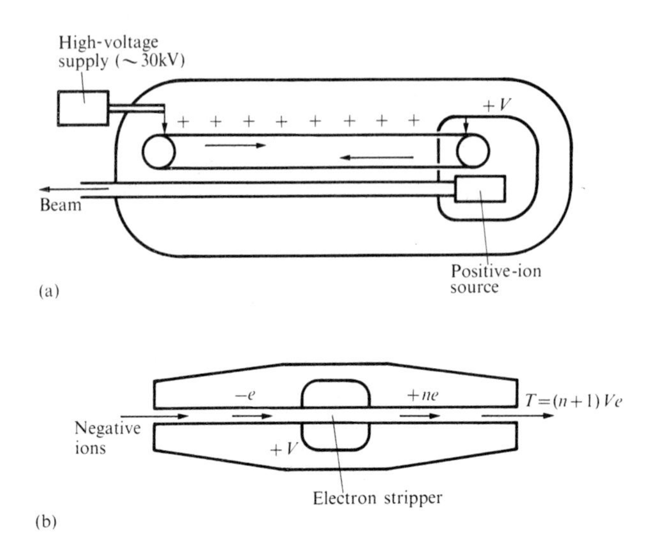

FIG. 4.2. (a) Schematic diagram of van de Graaff accelerator. V can be up to ≈ 12 MV. (b) Principle of tandem van de Graaff accelerator. V is maintained by a moving charged belt, as in (a).

containing a suitable gas, e.g. nitrogen with a few per cent of freon (CCl_2F_2).

In the tandem van de Graaff accelerator (Fig. 4.2(b)) a single terminal at high positive potential is used first to attract negative ions (e.g. H^-) from an ion source at earth potential and then, after a thin foil or gas has stripped them of two or more electrons, to repel the positive ions produced (e.g. protons) back to earth potential. Heavy positive ions can be multiply charged and the energies produced by a terminal at a potential V can therefore be $(n + 1)V$ eV if the negative ion is singly charged and the positive ion is a neutral atom less n electrons. The voltage is easily varied and can be very stable, making the machine ideal for many low-energy studies of nuclear structure and reactions.

The betatron is essentially a transformer with a single-turn secondary that consists of the electrons under acceleration circulating in an evacuated toroid ('ring doughnut'). The increasing magnetic flux produced by the primary produces a tangential electric field and also constrains the electrons to move in a circle. The condition for simultaneous acceleration and an orbit of constant radius R is simply that the instantaneous magnetic flux density at the radius R should be always one half of the mean flux density within the orbit. One bunch of electrons is accelerated in each period of a 50 or 60 Hz alternating current in the primary. A severe restriction on circular electron accelerators is the bremsstrahlung (radiation) loss which is $(1/4\pi\varepsilon_0)(4\pi e^2/3R)(E/m_0c^2)^4$ per turn. This is clearly negligible for heavier particles (see also pages 80 ff).

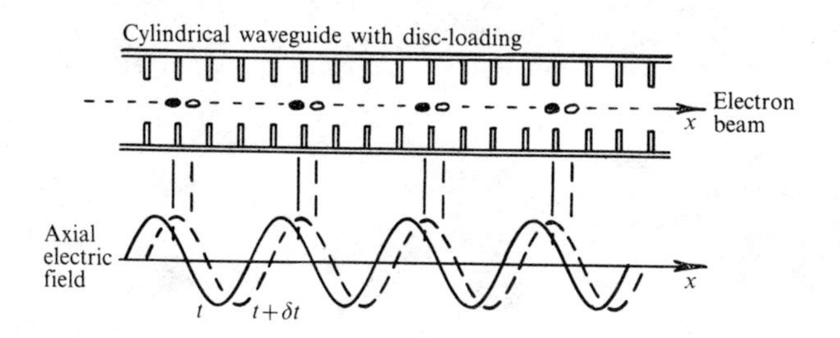

FIG. 4.3. Electron linear accelerator of travelling wave type. Bunches of electrons are shown solid at time t and hollow at $t + \delta t$.

The electron linear accelerator (Fig. 4.3) uses a cylindrical waveguide fed with radio-frequency power in a mode that provides a series of regions of positive and negative electric field along the axis. These field regions move along the guide at a velocity designed by means of 'disc-loading' to be that of a bunch of electrons under continuous acceleration in a positive-field region. The 20 GeV Stanford electron linear accelerator is two miles long.

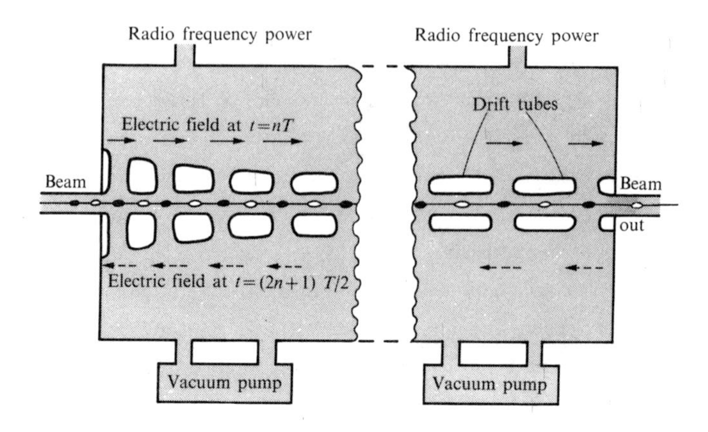

FIG. 4.4. Proton (or heavy ion) linear accelerator of resonant cavity type. T is the period of the applied radio frequency power. Bunches of particles are shown solid for the times of acceleration ($t = nT$) and hollow when they are shielded by the drift tubes.

REPEATED ACCELERATION

The proton linear accelerator acts on a different principle from that of the electron linear accelerator. A resonant cavity (Fig. 4.4) is used with a series of 'drift' tubes along the axis. The mode of oscillation of the injected radio-frequency power produces an electric field along the axis. At a given time the fields at all the gaps between the drift tubes are favourable for the acceleration of bunches of protons in the gaps. At unfavourable phases of the field the protons are within the metal drift tubes, and therefore shielded from the electric field. For $f = 200$ MHz, and $v = c/3$ ($\equiv$ 50 MeV protons) the distance between gaps is 0.5 m. The length of the machine for high energies soon becomes excessive and the power consumption is a problem. The radio-frequency power and the proton beam have to

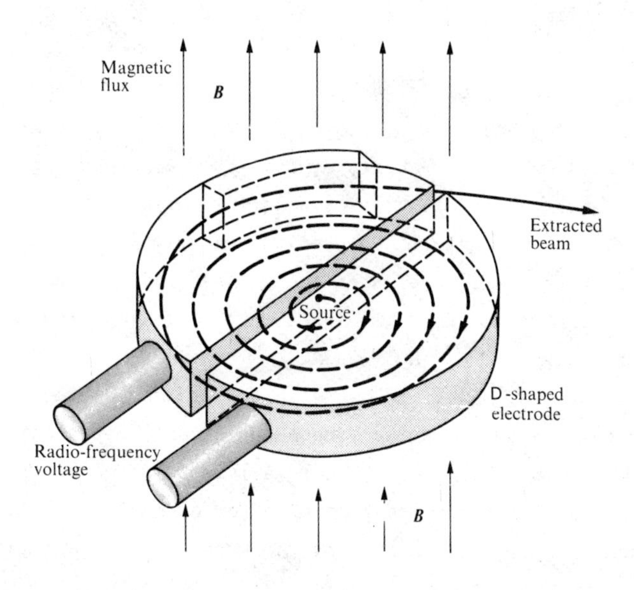

FIG. 4.5. Schematic diagram of simple cyclotron. The magnetic flux B is provided by a large electromagnet.

be pulsed so that they are 'on' for only ≃ 1% of the time. Superconducting resonant cavities with low radio-frequency power losses are being designed to improve this poor 1% 'duty cycle'. Proton linear accelerators usually have a Cockcroft-Walton injector and are themselves used as injectors for some synchrotrons (Fig. 4.7).

Orbital accelerators utilize the economy of space provided by the basically circular orbit arising from an applied magnetic field. The simple cyclotron (Fig. 4.5) relies on the fact that the period of revolution τ is independent of the velocity v of the particle. This follows from the familiar relation $qvB = Mv^2/R$. The period of revolution is $\tau = 2\pi R/v$ and therefore $\tau = 2\pi M/Bq$. Thus radio-frequency power supplied to the two D-shaped electrodes, known as dees, will produce at time $t = 0$ an accelerating field for a particle of mass M charge q at a given gap, and when $t = \tau/2$ it will produce acceleration at the diametrically opposite gap provided the applied frequency f is equal to the 'cyclotron frequency' $1/\tau = Bq/2\pi M$. Similarly there will be acceleration on its arrival at all subsequent gaps. When the electric field is unfavourable the particle is electrically shielded by the metal dees, but the magnetic field acts all the time. As the particle gains energy the orbit radius R increases until finally the particle is extracted at R_{max}, where the kinetic energy $T = q^2 R_{max}^2 B^2/2M$.

There is, however, a snag in the process at the higher energies since the particle mass is velocity-dependent, $M = M_o(1 - v^2/c^2)^{-\frac{1}{2}}$, owing to relativistic effects. Thus τ increases with v and the particle, for a machine with a large enough radius, would arrive at the gaps successively later than the times for optimum electric field until eventually the electric field at the time of arrival would be very small or zero.

At first sight the solution is obvious - increase B with R to keep in step with the increase of M with R. Unfortunately this would lead to forces with an axial component that would push the particles away from the median plane (Fig. 4.6(a)). In fact in the simple cyclotron B is made to decrease slightly with R in order to produce median-plane (or axial) focusing (Fig. 4.6(b)).

One way of overcoming the relativistic problem, which is used in the synchrocyclotron, is to concentrate on one bunch of particles at a time and gradually decrease the frequency of the applied radio-frequency power to allow the particles to arrive 'late' and yet still find an accelerating field. This turns out to be easier than it sounds because of an inherent phase stability whereby precise (and rapid!) orbit-by-orbit tracking of frequency and energy is not necessary. When one bunch of particles has been taken to the largest radius the frequency is increased to its original value and another bunch is accelerated.

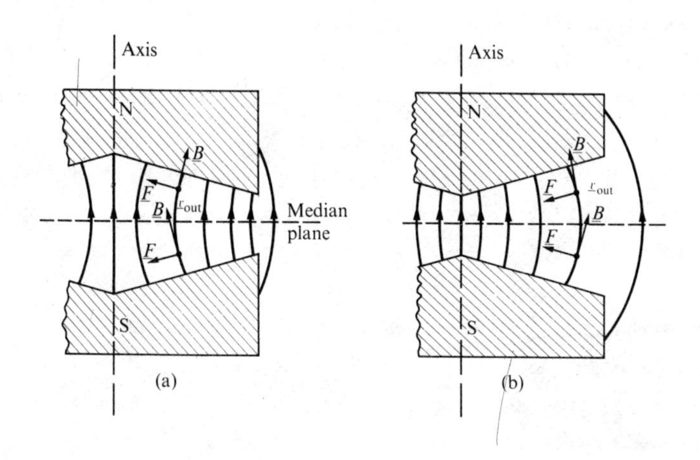

FIG. 4.6. (a) Defocusing effect of cyclotron magnetic field when B increases with radius. (b) Focusing effect when B decreases with radius. F is perpendicular to B and v. (Schematic and exaggerated).

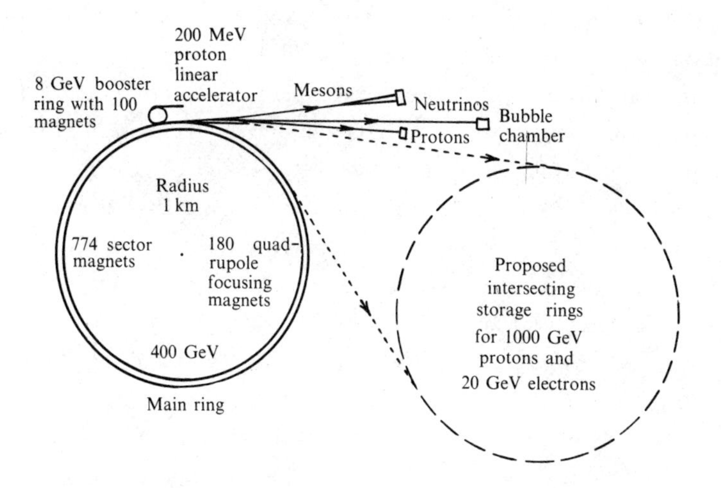

FIG. 4.7. Layout of proton synchrotron of the Fermi National Accelerator Laboratory, Batavia, U.S.A. with a proposed development.

The radio-frequency varies from about 30 MHz to 15 MHz at a modulation frequency of 50 or 60 Hz. The upper energy limit is set by the sheer size of the magnet which must produce a magnetic flux at all radii.

The synchrotron (Fig. 4.7) solves this new problem by using a fixed radius orbit and allowing the magnetic field acting on the orbit to increase as the particle momentum increases. Thus a ring magnet or series of magnets is used with a number of radio-frequency (RF) accelerating cavities arranged round the ring. The frequency still needs to decrease if protons are to be accelerated from an injection energy of (say) 50 MeV, at which energy $v \simeq c/3$, but for electrons at a few MeV injection energy the velocity is already very close to c. Thus a proton synchrotron normally needs to vary both B and f in unison, but for an electron synchrotron (or a proton synchrotron with an injector of several GeV) only B needs to be varied and f is constant. The limit is now purely financial, apart from bremsstrahlung losses for electrons, and for proton synchrotrons in the 400 GeV region the cost is at present approximately £1M per GeV. For conventional electromagnets covering 75% of the orbit, with B = 1.5 T, a proton energy of 400 GeV requires a radius of 1 km (Fig. 4.7). The development of superconducting magnets with B five or six times larger would enable the same radius machine to produce proportionately higher energies, i.e. more than 2000 GeV.

One of the important advances in the design of accelerators and beam transport systems was the application of the principle of strong alternating-gradient focusing. The simple axial focusing of Fig. 4.6(b) is achieved by using a variation of B with R of the form

$$B/B_o = (R_o/R)^n ,$$

with the field index $n \simeq \frac{1}{2}$ or less. This type of focusing, which is conveniently found to be associated with simultaneous focusing in the radial direction, is weak and it led to relatively large pole gaps in the early proton synchrotrons to accommodate diffuse beams; consequently it required very large magnets. If n is increased beyond unity, in practice to about 300, the effect is much stronger, and axial focusing is now found to be associated with radial defocusing. However, a focusing lens (convex, f_1 positive) in conventional optics, combined with a defocusing lens (concave, f_2 negative) distance d away, has a net focusing effect F given by $1/F = 1/f_1 + 1/f_2 - d/f_1f_2$. Hence, for $f_1 = |f_2| = f$, $F = f^2/d$. It follows that a combination of a sector magnet with positive n (B decreasing with R) and one with negative n (B increasing with R) produces focusing in both the axial and radial directions.

Alternating-gradient focusing is used in the proton synchrotron and has led to a considerable economy in magnet size because of the much smaller beam diameters (millimetres rather than centimetres). It is also applied in the sector-focusing cyclotron which has B increasing with R to compensate for the relativistic effect and, therefore, can use a fixed-frequency RF supply. The defocusing effect of the increase of B with R is more than compensated by the alternations of magnetic-field gradient produced by radial hills and valleys on the pole pieces, hill facing hill and valley facing valley. This machine yields essentially continuous currents of particles with quite high energies (approximately 600 MeV protons).

INTERSECTING STORAGE RINGS

A recent development has produced a remarkable leap upward in effective energy. By 'effective energy' we mean the energy available in the centre of mass (CM) of the particle system. This is an important concept in classical mechanics, but becomes of even greater significance in collisions involving relativistic velocities.

Consider a particle, mass M_1 and kinetic energy T_1, colliding head-on with a particle mass M_2 initially at rest, i.e. $T_2 = 0$. Conservation of linear momentum and energy in the non-relativistic case leads to a kinetic energy for the centre of mass of the resulting system given by $T_{of\ CM} = T_1M_1/(M_1 + M_2)$, and consequently the energy available within the system (i.e. in the centre-of-mass system) is $T_{in\ CM} = T_1M_2/(M_1 + M_2)$. For macroscopic colliding particles this energy in the centre-of-mass system could lead, for example, to a rise of temperature, emission of sound, or break-up. In nuclear collisions it can lead to excitation or break-up reactions. At sufficiently high energies new phenomena appear, notably the production of new particles such as pions (see pages 128 ff), but only if the energy in the centre of mass is greater than M_oc^2, where M_o is the rest mass of the new particles.

For most particle-production calculations it is necessary to use relativistic mechanics. In the simplest case of collisions between identical particles $M_1 = M_2$, and with M_2 at rest the result is $T_{in\ CM} = \{(T_1 + 2M_1c^2)2M_1c^2\}^{\frac{1}{2}} - 2M_1c^2$. In the high-energy limit $T_{in\ CM} \rightarrow (2T_1M_1c^2)^{\frac{1}{2}}$, and in the non-relativistic limit $T_{in\ CM} \rightarrow T_1/2$. Thus for proton-proton collisions with $T_1 = 30$ GeV (available at CERN, Geneva) and $T_2 = 0$, $T_{in\ CM} = 5.86$ GeV only.

For head-on collisions between identical particles of kinetic energy T there is zero momentum of the centre of mass, and all the original kinetic energy is available in the centre-of-mass system, i.e. $T_{in\ CM} = 2T$. Thus two 30 GeV protons colliding head-on yield $T_{in\ CM} = 60$ GeV. But in order for this to be available for a collision with a proton at rest the energy of the moving proton would need to be $T_1 = (T_{in\ CM} + 2M_1c^2)^2/2M_1c^2 - 2M_1c^2$, i.e. $T_1 = 2038$ GeV. For two 500 GeV protons, which could be made available from the Batavia (U.S.A.) synchrotron (Fig. 4.7), $T_{\text{in CM}} = 1000$ GeV, and for one proton at rest the other would need to have $T_1 = 862\,400$ GeV to produce this value of $T_{\text{in CM}}$!

The colliding-beam experiments with 30 GeV protons recently undertaken at CERN, Geneva, use 'intersecting storage rings'. Protons are 'stacked' in orbits in two highly evacuated rings which intersect at eight cross-over points. A switching system then allows the counter-rotating beams to collide. Significant new results are being obtained for energies equivalent to those of a 2000 GeV conventional machine.

Superconducting magnets may soon allow a proton synchrotron to be built to produce a 2500 GeV beam, and intersecting storage rings would then allow centre-of-mass energies of 5000 GeV. To produce 5000 GeV in the centre of mass using a sationary proton target would require an incident proton energy of 13 333 000 GeV = 1.333×10^{16} eV. Protons in cosmic rays have been observed with energies estimated to be as high as 10^{20} eV, but they are extremely rare, and of course unpredictable, occurrences.

PROBLEMS

4.1. Calculate the maximum energy of electrons accelerated in a betatron with a maximum magnetic flux density of 1.0 T at an orbit radius of 1.0 m.

4.2. What is the acting electric field, assumed constant, in the 20 GeV Stanford electron LINAC of length 2 miles? (1 mile = 1.61 km.)

4.3. The proton beam entering a linear accelerator has an energy of 300 keV. Calculate the distance between the first and second accelerating gaps for an applied radio-frequency field of 200 MHz, given that the first 50 gaps accelerate the beam to 10 MeV. Identify necessary assumptions and any approximations.

4.4. Calculate the ratio of the final to the initial radio-frequency for a 600 MeV proton synchrocyclotron, given that the necessary focusing is produced by a magnetic flux density at the final orbit which is 10% less than the value at the centre.

4.5. (a) A 50 MeV beam of protons is to be deflected through 90° by a uniform magnetic field $B = 1.0$ T acting perpendicular to the beam over a circular area. Calculate the minimum diameter of the magnet. (b) The same magnet is to be used to deflect the beam through 45°. Calculate the minimum value of B required. (c) Show that the exit beam has a direction along a radius of the magnet whenever the entry beam is initially directed towards the centre of the magnet.

4.6. Derive the formulae given on page 76 for the energy available in centre of mass for a particle of mass M_1, and laboratory kinetic energy T_1, incident on a stationary particle of mass M_2, (a) in the non-relativistic limit and (b) in the full relativistic form for $M_1 = M_2$. (Hint. For (b) use the invariance of $E^2-p^2c^2$, where p is the momentum, which is zero in the centre-of-mass (CM) system, and E is the total energy.)

5. Nuclear instruments and methods

ENERGY LOSS BY NUCLEAR PARTICLES

The design of instruments for the detection of nuclear and atomic particles requires an understanding of how the particles transfer energy to the detecting medium. The same understanding is needed in the design of shielding against the harmful effects, physical and biological, of exposure to radiation, and in the planning of medically beneficial exposures. Some applications of isotopes, e.g. the monitoring of the thickness of thin films during production, also depend on the characteristics of particle absorption.

Charged particles and photons interact with electrons in matter, and to a very much smaller extent with the nuclei. Neutrons interact with the nuclei by means of the strong nuclear interaction, but scarcely at all with the electrons. Neutrinos interact with the nuclei through the weak interaction only, and are very difficult to detect; the probability of interaction is so low that a neutrino produced during beta decay could traverse the diameter of the earth 10^{11} times with a 50% chance of survival.

Charged particles, other than electrons, lose energy at a rate shown in Fig. 5.1(a) and, except for the low-energy region, where multiply charged particles (e.g. alphas) capture electrons, the following equation holds

$$-dE/dx = (q^2 nZ/\beta^2) f_1(\beta, \mathcal{I}) \; ,$$

where qe is the charge of the particle, n the number of atoms (each with Z electrons) per unit volume of the absorber, and $\beta = v/c$. The function f_1 of β and of $\mathcal{I}$, an appropriate mean

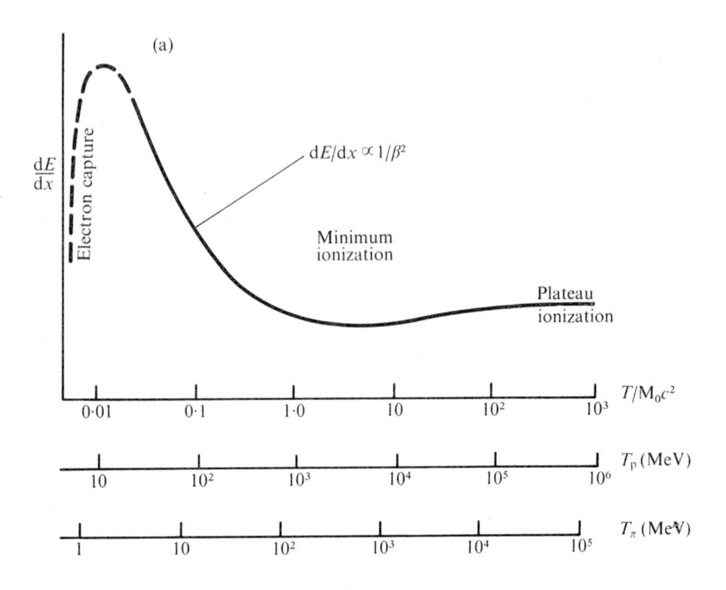

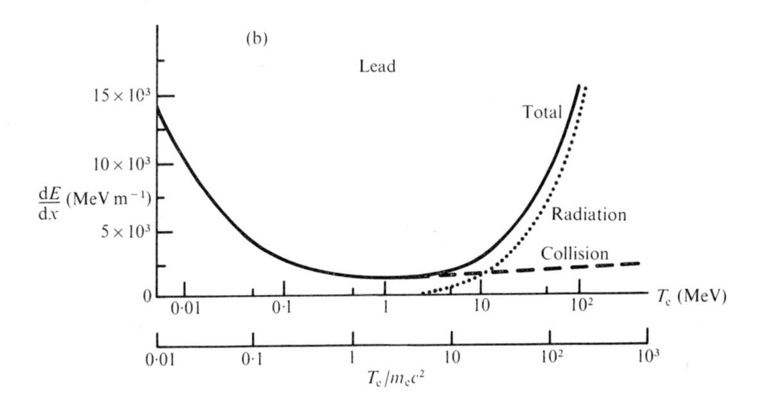

FIG. 5.1. (a) Variation of the rate of energy loss, dE/dx, with the kinetic energy (T) for 'heavy' charged particles (i.e. other than electrons) of rest mass M_0. Kinetic energy scales are shown for protons and pions. (b) Rates of energy loss for electrons in lead, showing the contributions from bremsstrahlung (Radiation) and ionization and excitation (Collision).

of the excitation and ionization energies ($\mathcal{I}$ = 13.5Z eV), is essentially constant for kinetic energies less than $0.2M_0c^2$, where M_0 is the rest mass of the incident particle, for which region therefore dE/dx is proportional to $1/\beta^2$. Notice that dE/dx expressed as a function of β is, for a given absorber, independent of the mass M_0 of the charged particle. But the range R of a particle, obtained by integrating dE/dx, does depend on M_0, since the appropriate variable is β and $dE = M_0c^2\beta d\beta$ (non-relativistic). For $\simeq$ 10 MeV < T < $\simeq 0.2M_0c^2$ we can approximate dx is proportional to $(M_0\beta^3/q^2nZ)d\beta$, i.e. R is proportional to T^2/M_0q^2nZ. If we express R in kg m^{-2} (rather than metres), we need to multiply both sides of this proportionality by the density ρ of the absorber, but we note that $n = N_A\rho/A$, and very approximately A is proportional to Z, i.e. R (kg m^{-2}) is proportional to T^2/M_0q^2 and is approximately independent of the absorber material. A corresponding approximate empirical relation is $R = 0.034\ T^{1.8}/M_0q^2$ (kg m^{-2}), for T in MeV and M_0 expressed in proton masses. Notice the close parallel to Geiger's relation (see page 24), expressed as T_α (MeV) = $2.12\ \{R\ (\text{cm of air})\}^{\frac{2}{3}}$.

Ionizing particles can impart considerable energy to struck electrons, and these secondary electrons, known as delta rays, can themselves produce ion pairs and show their paths in suitable track detectors. When the primary particle is moving at high relativistic speeds, the number of delta rays per unit length of primary track is simply proportional to q^2, and affords a means of particle identification. For large q values the end of the track has a diminishing dE/dx, because the particle begins to capture electrons and reduce its effective q (Fig. 5.1(a)), and this diminution can also be used to assess q. These measurements have been of considerable value in studies of the charge spectrum (and hence the mass spectrum) of high-energy primary cosmic rays, using nuclear emulsion detectors.

Energy loss by electrons involves not only excitation and ionization of atoms but also bremsstrahlung or 'braking radiation' (Fig. 5.1(b)). Whenever a charged particle is accelerated or decelerated it emits electromagnetic radiation. Multiple deflections of fast electrons by the Coulomb fields of nuclei in an absorber involve changes of momentum, and therefore involve accelerations. (It is the periodic accelerations of electrons in radio transmitting aerials that lead to radio-wave emission.) The rate of emission is given by the classical equation

$$-(dE/dt)_{rad} = (1/4\pi\varepsilon_o)(2e^2/3c^3)|d^2x/dt^2|^2.$$

For a particle of charge qe, mass M_o, approaching a nucleus of charge Ze, the force $M_o(d^2x/dt^2)$ is proportional to qZ/x^2, and hence $-(dE/dt)_{rad}$ is proportional to $q^2Z^2/M_o{}^2$. Clearly the inverse dependence on $M_o{}^2$ makes the bremsstrahlung energy loss for protons quite negligible compared with its value for electrons. A complete analysis yields approximately $-(dE/dx)_{rad} \propto q^2Z^2T/M_o{}^2$ and for electrons this is the main term in the energy loss for $T > \sim 10$ MeV for heavy elements (e.g. lead, Fig. 5.1(b)) and for $T > \sim 100$ MeV for carbon.

Neutrons lose energy predominantly by elastic collisions with nuclei. At non-relativistic energies it is easy to show that the maximum energy transferred for a head-on collision of a neutron, initial energy T_o, with a nucleus of mass number A is

$$\Delta T = 4AT_o/(A + 1)^2 ,$$

hence materials of small A are more effective than those of large A. This is important in the choice of 'moderators' for nuclear reactors (see page 114). Inelastic collisions also take place, and neutron absorption can be very important for $T_n < \simeq 1$ MeV.

Energy loss by gammas is possible by three main processes (1) photoelectric effect, (2) Compton effect, and (3) pair production. In the first the photon disappears, and its energy $h\nu$ is used to free an electron and give it kinetic energy. In the Compton effect the photon is scattered by an (almost) unbound electron, mass m_o, and the change in wavelength and hence the energy loss is related to the scattering angle θ by $\lambda_1 - \lambda_o = (h/m_oc)(1 - \cos\theta)$. The electron and positron (see pages 144 ff) from pair production share the kinetic energy $h\nu - 2m_oc^2$, and when later the positron annihilates with an electron at rest two photons are most commonly produced each of energy $h\nu_1 = m_oc^2 = 0.51$ MeV. In all three processes energetic electrons are emitted and lose energy to the absorber by excitation, ionization, and possibly bremsstrahlung - the last being, of course, photons of (much) lower energy than the primary gamma photon.

The relative importance of these processes as a function of photon energy, and how they differ for aluminium and lead, can be seen in Fig. 5.2. The intensity I_x of the radiation transmitted through a thickness x is given by $I_x = I_o \exp(-\mu x)$ where I_o is the incident intensity and μ is the linear absorption coefficient. It is important to note that I_x is the intensity of radiation with the same photon energy $h\nu$ as that of I_o, and there is always a 'build up factor' to be considered in designing shielding and computing medical dosages to allow for the build up of degraded photons of lower energy from the Compton effect and positron annihilation.

The total energy absorbed by a medium exposed to ionizing radiation is termed the absorbed dose and the unit (1953) is the rad (10^{-2} J kg^{-1} = 6.25×10^{10} MeV kg^{-1}). Recently the unit gray was approved (1Gy = 100 rad). For living organisms damage to tissues arises mainly from molecular disruption in individual cells in the track of the particle. The same dose for particles

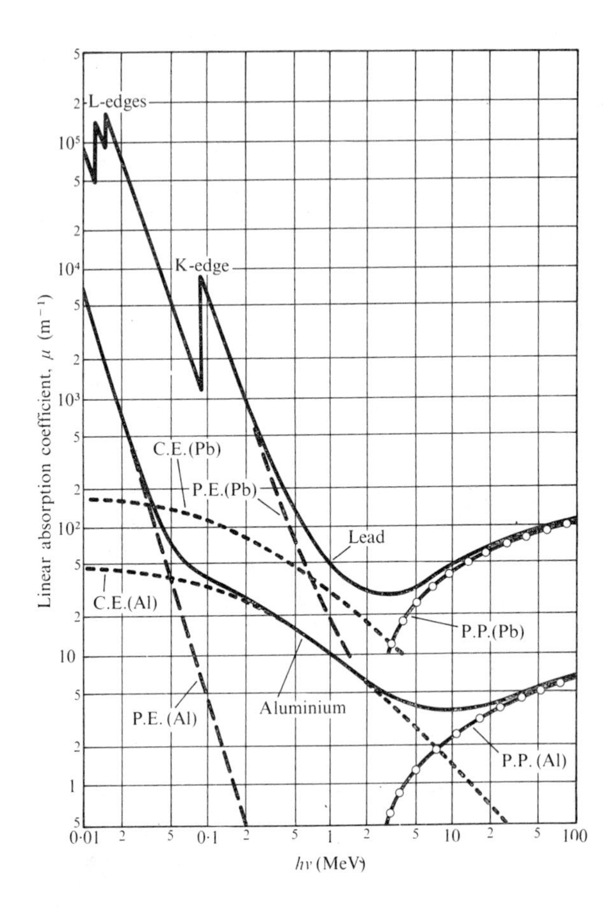

FIG. 5.2. Linear absorption coefficients for X-rays and γ-rays in aluminium and lead. The total coefficient μ is the sum of the three contributions; photo-electric effect (P.E.), Compton effect (C.E.) and pair-production (P.P.) Notice that both scales are logarithmic.

with characteristically different types of tracks produces different biological effects and a factor known as 'relative biological effectiveness' (RBE) is introduced and used to define a 'dose-equivalent' (DE), measured in rems (rem derives from 'roentgen equivalent man' - see later); DE(rem) = dose(rad) × RBE.

Thus RBE is 1 for gammas and electrons, 10 for alphas, and 5-10 for fast neutrons. The earliest unit for radiation exposure was proposed by Villard in 1908 and came to be known as the roentgen. It is not very convenient as it refers only to the ionization produced when X-ray or gamma photons are absorbed in air. An exposure of one roentgen (R) produces 2.58×10^{-4} C kg^{-1} of ions, and since an average of 32.5 eV is required to produce a pair of ions, one roentgen deposits 0.838×10^{-2} J kg^{-1}, and is approximately equivalent to a dose of one rad.

For workers regularly exposed to nuclear and X-radiation, and therefore subject to statutory regular medical supervision, the maximum permissible radiation dose, or 'tolerance' dose, for whole-body radiation, has been successively reduced from 72 rem per year in 1928, to 30 rem per year in 1940, to 15 rem per year in 1945, and since 1959 it has been 5 rem per year. For comparison it should be noted that cosmic rays at sea-level produce approximately 1% of the permitted dose. An individual member of the general public is permitted a maximum dose of only 0.5 rem per year, whilst the average maximum permitted dose for the whole population is set at 0.17 rem per year. For a single whole body dose of 400 rem there is a 50% chance of death; for 600 rem death is almost certain.

PARTICLE DETECTORS

Most particle detectors utilize the electrons or positive ions produced in the detecting medium to generate an electric pulse. In the gas ionization counter the electric field between two plates or grids collects all the primary ions, both positive and negative. The resultant pulse for one incident particle is very small, but the detector is more often used for integrating the pulses from a flux of X-rays or gamma rays to produce a small, steady current.

The gas proportional counter uses higher voltages between

the electrodes, and the primary ions gain sufficient energy to produce secondary ion pairs. Multiplications up to 10^6 are obtained for electric fields of a few hundred volts per centimetre, and since the output pulse is proportional to the initial number of ions useful information is obtained about the total energy of the incident particle, if it stops, or about its dE/dx and hence its identity if only part of its track intersects the sensitive region.

The Geiger (or Geiger-Mueller) counter (see page 22) uses even higher electric fields, and the release of ultraviolet photons by the numerous ions propagates the discharge throughout the counter volume, so that the size of the output pulse, which is large, is independent of the initial number of ions. It is a sensitive, robust, and useful instrument for simple monitoring of ionizing radiation, but by itself it gives no information on the identity or energy of the detected particle.

Pocket dosimeters (Fig. 5.3) are gas ionization counters that indicate the leakage of high voltage from an insulated electrode, rather like a gold-leaf electroscope. The total movement of a gold-plated flexible quartz fibre can be viewed against a scale by means of an eyepiece, and hence the integrated exposure estimated directly in millirads (or milliroentgens).

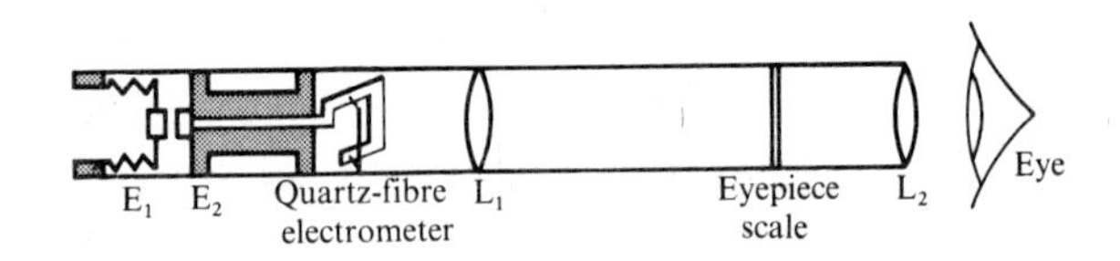

FIG. 5.3. Schematic view of pocket dosimeter (≃ 12 cm long). The lens L_1 forms an image of the quartz fibre on the eyepiece scale which is viewed through the lens L_2. The spring-loaded electrode E_1 allows recharging of the insulated electrometer by contact with E_2.

Solid ionization counters, consisting of a suitably treated semiconducting crystal (germanium or silicon), find increasing applications because solids are about 1000 times more dense than gases, and hence more compact, and furthermore they require much less energy to produce an ion pair ($\simeq$ 3 eV compared with $\simeq$ 30 eV for gases). The increased numbers of ions are of importance not so much for the larger pulses produced, since amplification is always needed, but because the mean percentage statistical variation (100 $\delta N/N$ is proportional to $N^{-\frac{1}{2}}$) in the pulse size for a given energy deposited is significantly reduced, yielding excellent energy (or dE/dx) resolution. For low-energy particles, and for gamma-ray spectrometry, a germanium crystal (maximum volume $\simeq$100 ml) with lithium diffused into it, to produce a carrier-free region, is an excellent detector except for the need to use and store it at liquid-nitrogen temperatures. An example of its resolution is shown in Fig. 5.4. Other types of semiconductor detector, the silicon surface-barrier and the n-p

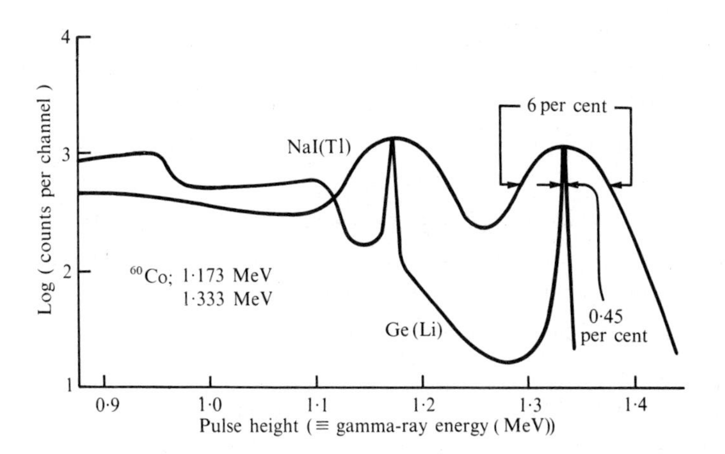

FIG. 5.4. Comparison of resolution achieved by NaI($T\ell$) scintillation counter (3" x 3" cylinder) and Ge(Li) semiconductor detector for ^{60}Co gamma rays. Note the two Compton scattering edges near 0.96 and 1.11 MeV in the Ge(Li) spectrum.

junction detectors, have relatively shallow (≃ 1 mm) sensitive regions, but have the advantage that they operate at room temperature.

The present-day scintillation counter is a striking example of technological development of a simple device. In 1903, Crookes and, independently, Elster and Geitel discovered that the irradiation of zinc sulphide by alphas produced individual flashes of light rather than the expected steady luminescence. In this early work, a microscope and a dark-adapted eye were essential, but these have been replaced by an electron-multiplier, in which the photons are first converted to photo-electrons and then these electrons are multiplied approximately three-fold by accelerating them through 100 V to a specially prepared 'dynode' that yields secondary electrons. The multiplication process is commonly repeated by ten more stages ($3^{10} \simeq 6 \times 10^4$) to yield useful electric pulses at the final collecting electrode or anode. The multiplication is roughly proportional to V^6, where V is the total voltage (≃ 1 kV) across the dynodes, and therefore highly stabilized extra-high-tension (EHT) supplies are required for high-resolution analyses of pulse heights. To a good approximation the size of the pulse output is proportional to the light input, and this is proportional to the energy deposited by the nuclear particle or gamma photon in the scintillator. The most common scintillator for gammas is thallium-activated sodium iodide (NaI(Tℓ)) which has a rather slow release of the scintillation photons (time constant 0.25 μs). Plastic scintillators (and liquids) are available with time constants of a few nanoseconds, and are used extensively in low- and high-energy research. The obtainable resolution is not so good as that of semiconductor detectors, but very large volumes of plastics with linear dimensions up to ≃ 1 m are available. Relatively large (≃ 2 litre) NaI(Tℓ) crystals, which are deliquescent and need to be encapsulated, have been produced for

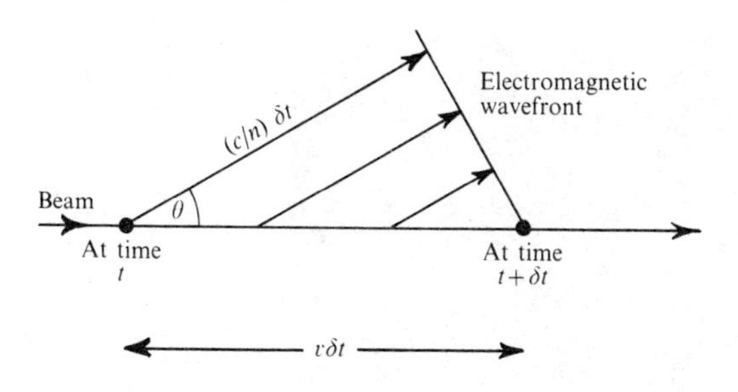

FIG. 5.5. Čerenkov effect. For particle velocity v greater than the phase velocity c/n, light is emitted parallel to the (generator of the) surface of a cone of half-angle θ.

high gamma-detection efficiency. An example of the resolution for NaI(Tℓ) is shown in Fig. 5.4.

Photomultipliers are also used in Čerenkov detectors which depend on the electromagnetic radiation, mostly visible or ultraviolet, produced by a charged particle when it enters a medium with a velocity that is greater than the (phase) velocity c/n of light in the medium, where n is the refractive index of the medium. The intensity is small but the direction of emission of the light (Fig. 5.5) is at a characteristic angle θ to the track, given by $\cos\theta = (c/n)\delta t/v\delta t = 1/\beta n$. It is sometimes very useful to be able to make a direct measurement of θ and hence obtain the velocity of a particle. The Čerenkov effect is the optical analogue of the sonic boom.

TRACK DETECTORS

There is no doubt that a picture (Fig. 5.6) of the tracks of a nuclear event or 'explosion' carries more conviction for many people than complicated deductions based on pulse heights, pulse shapes, and counting rates from particle detectors. It is true that the time relationships, and even the sequence of events, are usually lost, but much information is recorded. Thus the density

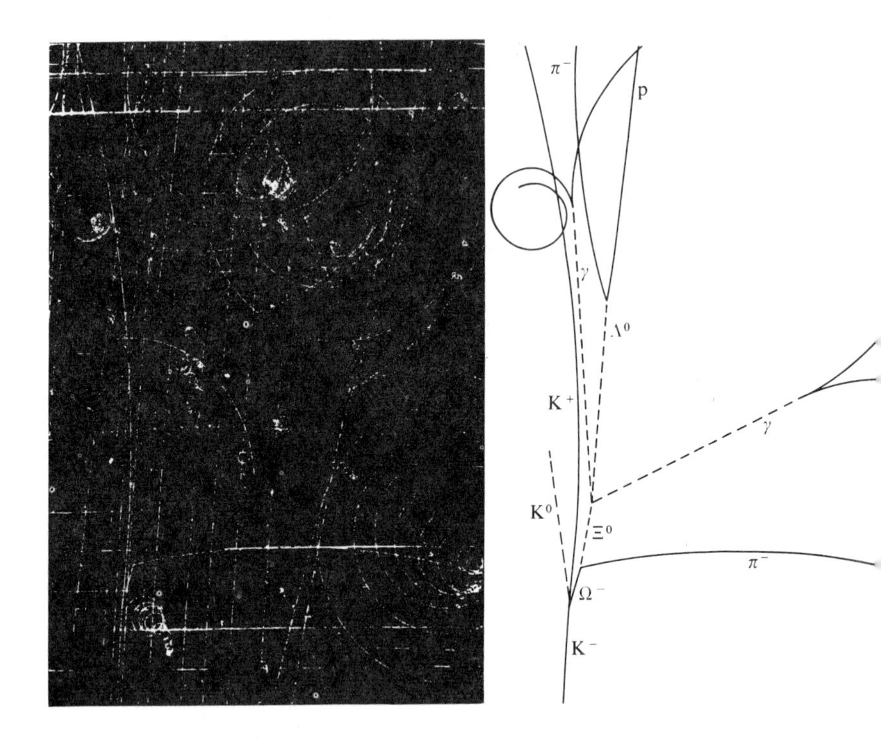

FIG. 5.6. Photograph from liquid hydrogen bubble chamber with a (perpendicular) magnetic field and the interpretation of the event of interest, which is the historic discovery of the omega-minus (see pages 165 ff). The dashed lines are reconstructions of the trackless neutral particles. The impressive spirals are from the electron-positron pairs produced by each γ-photon, and are explained by the steady loss of energy of the electron and the consequent steady reduction of track radius. (V. E. Barnes and 32 others, Phys. Rev. Lett. 12, 204 (1964).)

of the track can yield dE/dx and the delta ray density can indicate charge; the range relates to the energy, mass, and charge, and the angles give linear momentum relations. If a strong magnetic field is used there is the added information (momentum/charge) provided by the curvature of the track.

Although uncharged particles leave no tracks they can produce associated charged particle events and these, together with 'missing mass' and 'missing momentum' considerations, can lead to the identification of the uncharged particles.

The earliest track detector was the Wilson cloud chamber (1912), in which the positions of ions produced along the track of the charged particle in a gas are made visible by causing water droplets to form upon them (Fig. 8.2). This is achieved by allowing the particle to enter a chamber containing a gas saturated with water vapour and then suddenly expanding the gas by moving the piston-like floor (or ceiling or wall) of the chamber. The associated decrease of gas temperature produces supersaturation, and the ions act as favoured centres for the growth of drops. Stereoscopic flashlight photographs enable the three-dimensional characteristics of events to be measured. The diffusion-type cloud chamber has continuous sensitivity in a horizontal layer that is made supersaturated by an applied temperature gradient. Warm vapour diffuses downwards towards the base, which is usually cooled with solid carbon dioxide (dry ice).

The bubble chamber, which utilizes the local heating along the track of a charged particle in a liquid, is particularly useful in conjunction with accelerators that produce pulses of particles at predetermined times. About 15 ms before the particles enter, a signal from the accelerator causes the pressure in the chamber to be suddenly reduced, in order to provide a superheated liquid. The local heating along the track produces bubbles which are photographed before the chamber is quickly recompressed to prevent boiling throughout the volume. Notice that it is not possible to 'trigger' the bubble chamber by using a signal (e.g. a scintillation) from the entering particle to actuate the pressure release. During the few milliseconds needed to effect the decompression the local heating would be dispersed and the track would disappear. Enormous bubble chambers have been built with dimensions of several metres (5 m at Batavia, U.S.A.), usually with liquid hydrogen, since the detecting medium is of course also the nuclear target for reactions and proton targets are of fundamental importance. Some chambers contain organic liquids or other higher-Z liquids in order to improve gamma-detection efficiency. The use of magnetic fields for determining momentum/charge, and also the sign of the charge, has been improved by the development of superconducting magnets built around the bubble chamber.

Special nuclear photographic emulsions had their hey-day in the period 1945-65 in relation to research in cosmic rays and elementary particles. Large volumes (≃ 0.3 m cube) of highly concentrated ultrafine-grain photographic emulsions were produced by stripping ≃ 0.5 mm thick emulsions from their glass plate backing and stacking them with interleaving lens-tissue. For the tedious developing or processing they were first remounted on specially prepared glass to prevent undue distortion. Examination of nuclear emulsion tracks requires high-power (≃ × 2000) microscopes and large teams of observers, although some automatic scanning devices have been used.

Several other types of detectors have been based on the conversion of tracks of ions in a gas into a series of sparks between successive plate electrodes pulsed to sufficiently high potential differences. In the simplest form of 'spark chamber' the spark tracks are photographed, but the delay and lack of precision in analysis are frustrating. In an improved version the acoustic pulse emitted by a spark is picked up by three or more microphones, and the times of arrival of the signals fed straight to a computer for 'on-line' analysis of the position of the spark.

The most extensive applications of 'instant' track detectors have been based on wire chambers. A planar array of very thin wires, covering areas of several square metres, is used to produce sparks between adjacent wires, or alternatively a pulse of current from the primary ions after a proportional-counter type of amplification. The current pulses in the wires are monitored by individual electronic devices that enable almost simultaneous analyses to be undertaken by a computer. These gas ionization detectors can be triggered either by a pulse from the accelerator that produces the primary particle beam or by scintillation counters on each side of the array, provided that their signals are in time-coincidence. The use of such auxiliary detectors allows much more efficient experiments to be designed. Even so, one of the problems of nuclear research is the recording and analysis of vast amounts of experimental data.

ISOTOPES

Distinct nuclear species are commonly referred to as 'isotopes', whether radioactive or stable. Strictly speaking, isotopes are of the same element, i.e. in the same place (Greek: topos) in the classification table of the elements, and therefore have the same number of protons. Similarly 'isotones' are nuclei with the same number of neutrons, and 'isobars' are nuclei with the same atomic mass number A. Distinct nuclear species are properly termed 'nuclides' (or sometimes nucleides), but we shall mostly use the popular term isotopes. Nuclear 'isomers' are excited nuclei with lifetimes greater than about 1 μs. ^{137}Ba (2.7 min) and ^{110}Ag (253 days) are examples of the relatively few that have lifetimes longer than 1 s. An isomeric transition is the emission of a gamma from an isomer or isomeric state.

Isotopes were initially recognized in the three natural radioactive series and later identified by means of mass spectrometers, first by Thomson and Aston (1921) for the neon-20, 22 isotopes. There are about 280 'stable' isotopes found in nature, including a few such as ^{40}K (0.0118% of natural potassium) with a half-life of 1.3×10^9 years (β^-, e capture),

^{204}Pb (1.48%, 1.4 × 10^{17} years α) and ^{238}U (99.27%, 4.51 × 10^{9} years α). Within the naturally occurring radioactive series there are some 50 unstable isotopes, and as a result of nuclear-physics research many hundreds of artificial radioactive isotopes have been discovered, ranging from tritium (^{3}H, 12.26 years, β^-, 18 keV) right up to the new transuranic elements with atomic numbers 93 - 106 (see page 107, and Table 6.1). Some of these, and some of the rarer stable isotopes, have found many applications in medicine and industry, as well as in chemistry, geology, and biology research. As a result methods of preparation and separation have been developed in specialized supply laboratories.

The major sources of radioactive isotopes are (1) waste products from nuclear reactors, (2) irradiation of stable isotopes with neutrons within reactors, and (3) irradiation with particles from accelerators. Only longer-lived isotopes, e.g. ^{90}Sr(28 years) and ^{137}Cs(30 years), are usefully extracted from spent reactor fuel rods because of the tedious chemical separations involved. A large number of useful isotopes can be made by irradiation with neutrons in a reactor, and in nearly every case the radioactive product is of the same chemical element, e.g. ^{24}Na(15 hours, β^-) from ^{23}Na(100%), and ^{60}Co(5.26 years, β^-) from ^{59}Co(100%). This makes chemical separation of the radioactive isotope almost impossible, and therefore the specific activity (Ci kg^{-1}) is often rather low. Szilard, and also Chalmers, showed that if the irradiated element is in a chemical compound the gamma emitted after neutron capture causes a recoil that breaks the chemical bond. Free radioactive atoms are formed which readily allow chemical separation from the compound, e.g. ^{128}I(25 min, β^-) is formed from ^{127}I(100%) in ethyl iodide. Deuterons are the commonest accelerated particles used in isotope production but the yields of isotopes are rather small, perhaps at most a few curies of activity. On the other hand, the product can be chemically dissimilar from the initial element and the easy separation can yield high specific activity, e.g. ^{22}Na (2.6 years, β^+, e capture), which is the longest-lived positron emitter, from $^{24}Mg\ (d,\alpha)^{22}Na$.

The separation of stable isotopes can be effected by a wide variety of methods, but all are expensive and inefficient. Magnetic separators give good mass resolution, but the limitations of their ion sources lead to very small yields, and the method is confined to the more exotic isotopes, e.g. new nuclides. Gaseous diffusion was the method selected, in the development of the nuclear bomb, for the separation of ^{235}U(0.72%) from ^{238}U(99.28%) in the form of the gas uranium hexafluoride (UF_6), there being only one stable isotope of fluorine (^{19}F). Strictly speaking, it was 'effusion' through the minute holes ($\simeq 10^{-8}$ m diameter) in thin sheets of a silver-zinc alloy, etched by hydrochloric acid. The rate is proportional to $1/M^{\frac{1}{2}}$, and some 4000 stages are necessary for 99% ^{235}U. In thermal diffusion, using liquids or gases, lighter molecules move to regions of higher temperature, e.g. near a hot wire on the axis of a long (up to 35 m) vertical cylindrical tube. The method is useful for laboratory scale work. Recently centrifugation has been tried with good results.

The separation of deuterium (^2H, 0.015%) from natural hydrogen is relatively easy and has great potential importance in relation to nuclear fusion reactors (see page 116). The best large-scale process utilizes chemical exchange, a principle which can also be used for other low-A elements. Natural hydrogen gas and steam are passed up a vertical tube containing a catalyst. Cold water is passed down and by chemical exchange the deuterium from the natural hydrogen concentrates in this water. The enriched water is used to provide the gas and steam for the next stage. An alternative method is repeated electrolysis of water leading to concentration of deuterium in the residual water.

Applications of isotopes are many and widespread. In medicine their uses include both diagnosis and therapy. Thus a measure of the rate and location of the highly preferential uptake of $^{131}I(\beta^-,\gamma,8.1$ days) by the thyroid and of the rate of its excretion in the urine enables abnormal thyroid action to be studied. In cases of excess thyroid hormones in the blood (hyperthyroidism), a dose of ^{131}I ($\simeq$ 5 mCi) yields relief in some 75% of cases by virtue of the damage to the thyroid cells produced by the betas from the locally absorbed isotope. Some forms of cancer are treated by localized application of isotopes (e.g. ^{137}Cs (β^-,γ,27 years) in the uterus), and the development of powerful collimated $^{60}Co(\beta^-,\gamma$,1.17 MeV, 1.33 MeV, 5.2 years) gamma sources ($\simeq 10^6$ Ci) provides for general radiation therapy the modern equivalent of the lower energy X-ray machines.

In industry ^{60}Co gamma sources are used in the radiography of weldings and castings. Sterilization of food products, drugs, and medical instruments is sometimes undertaken, but massive doses (10^6 rad) are required. The thickness of plastic and metal sheets can be monitored, during production, and thus controlled through feedback systems, by count-rates of transmitted or backscattered ($\theta \simeq 180^\circ$) beta particles. Transmitted alphas and gammas are also used, depending on the thicknesses and materials involved. Leaks and blockages in underground pipes (e.g. oil pipes) can be traced by the introduction of radioisotopes. Other applications overlap the use of 'tracers' in chemical and biological research, i.e. the labelling of certain molecules with a radioactive isotope which, in general, has the same chemical properties as its stable equivalent. Thus the wear on a radioactively doped or irradiated piston can be estimated from the radioactive build up in the lubricating oil, and rates of diffusion in metals and plastics can be measured even though the amounts of material involved are exceptionally small.

In organic chemistry and biology, ^{14}C (β^-, 155 keV, 5600 years) has many applications as a tracer, e.g. in exchange and transfer reactions. Owing to its continuous and apparently steady production in the atmosphere by cosmic rays, living plants and animals have very closely the same fraction of their carbon as ^{14}C, about 1 in 8×10^{11} atoms, yielding 14 β^- disintegrations per gram of carbon per minute. At death the exchange with the atmosphere ceases and radioactive decay sets in. Subsequent measurements allow the elapsed time to be estimated and dating of woods, for example, can be made for ages up to 50 000 years. Oxygen and nitrogen have radioactive isotopes with half-lives of seconds or minutes, much too short for general applications. It is then sometimes convenient to use stable but rare isotopes of these elements and trace their involvement in processes by mass spectrographic analyses. Another method is to activate the tracer by neutron bombardment in a reactor, a variant of 'activation analysis' of trace elements in geological and other specimens which depends on careful gamma-spectrographic analysis using, for example, Ge(Li) solid-state detectors.

The range of applications of isotopes is already very great, but it is certain that the possibilities are far from exhausted.

PROBLEMS

5.1. (a) What is the energy of a proton that has half the range of a deuteron of energy T? (b) What is the relation between the range and kinetic energy for particles with the same initial dE/dx?

5.2. Derive an expression for the energy lost by a neutron of initial energy T_0 in scattering at an angle θ by a nucleus of mass A. (Hint. Solve first in the centre-of-mass system and then relate $\cos\theta$ and $\cos\varphi$, where φ is the scattering angle in the centre of mass.)

5.3. Calculate the maximum energy imparted to an electron in

Compton scattering of photons of energy 1.173 MeV and 1.333 MeV (see Fig. 5.4).

5.4. Use Fig. 5.2 to estimate the relative thicknesses of lead and aluminium to produce a given attenuation for photons of energies 10 keV, 100 keV, 1 MeV, 10 MeV, 100 MeV.

5.5. A Geiger-Mueller counter has associated electronic circuits that 'paralyse' the counter for known adjustable times τ after each registered count in order to allow the ionization in the counter to be returned to zero.

(a) Show that the count-rate N_c after correction for the paralysis time is given by $N_c = N_o/(1-N_o\tau)$, where N_o is the observed count-rate.

(b) N_o is recorded for a range of values of $\tau = 300 - 1200\ \mu s$ and $N_c = 2000\ s^{-1}$. Present the results as a straight-line graph, and deduce what you can from the slope and intercepts.

5.6. A 5 MeV alpha particle is stopped in a gas ionization counter. Assume that 30 eV is required to produce an ion pair and that all ions are collected, what will be the change in voltage if the capacitance of the counter is 50 pF?

5.7. Calculate the rest mass of the particles that begin to emit Čerenkov radiation in a medium of refractive index 1.5 when their kinetic energy exceeds 37 MeV (see Table 8.1).

5.8. A solution containing 0.05 μCi of ^{24}Na was injected into a patient's bloodstream, and after 5 hours the activity in 10 cm^3 of the patient's blood was found to be 140 min^{-1}. Calculate the total blood volume of the patient.
(^{24}Na $\tau_{\frac{1}{2}}$ = 15 hours.)

5.9. Radiocarbon-dating assumes that the ratio of ^{12}C to ^{14}C in the Earth's atmosphere has remained constant at (about) 8×10^{11} during the periods in question. A sample of ancient wood is reduced to pure carbon, 1 g of which has an activity of 3.0 pCi. Determine the age of the wood. (^{14}C $\tau_{\frac{1}{2}}$ = 5568 years.)

5.10. A sphere of plutonium - 239, of mass 0.1 kg, is suspended in an enclosure at a temperature of 288 K. Given that its half-life is 24 400 years, and that the total disintegration energy in alpha emission is 5.24 MeV, calculate (a) its activity and (b) its equilibrium temperature, noting the errors introduced by any simplifying assumptions. (Density of ^{239}Pu = 19 000 kg m^{-3}, Stefan-Boltzmann constant = 5.67×10^{-8} J s^{-1} m^{-2} K^{-4}.)

5.11. A foil of aluminium - 27, 0.02 mm thick, is bombarded by a 0.5 mA beam of 1 MeV protons. A proton-alpha reaction takes place with a cross-section of 20 barns. How long must the foil be bombarded in order to produce two micrograms of the stable product element?

5.12. Show that the activity of a foil containing N nuclei, after irradiation for a time t by a flux φ of nuclear particles is given by

$$A = \varphi\sigma N(1-\exp(-\lambda t))$$

where σ is the cross-section for formation of the product of radioactive constant λ, assumed to decay to a stable end product.

6. Nuclear reactions

INTRODUCTION

Nuclear reactions, in the widest sense of this term, include (1) elastic scattering, (2) inelastic scattering, and (3) processes in which there is a change of identity in one or more of the nuclear particles involved (Fig. 3.1). Sometimes the term nuclear reaction is restricted to category (3), and sometimes to categories (2) and (3). These distinctions are usually obvious in any given context but need more careful consideration in relation to the 'total' cross-section (see pages 104 ff).

Several examples of reactions have already been introduced, e.g. in Chapter 3, where the need to conserve mass-energy was emphasized, and in Chapter 4 where the importance of centre-of-mass energy was introduced and the need to allow for relativistic effects was illustrated. In this chapter we shall develop some of these principles and give examples of nuclear reactions that have contributed to a deeper understanding of nuclear physics or have led to important applications, such as nuclear reactors for energy production.

ENERGETICS

Consideration of mass-energy conservation yields information on the energetics of a reaction. If the sum of the rest masses of the final particles is less than the sum of the initial masses the reaction is said to be 'exothermic', i.e. literally 'gives out heat', a rather misleading phrase.

For the reaction A(a,b)B, i.e. particle a on target A producing particles b (light) and B (heavy), we can write in terms of <u>nuclear</u> masses and the Q-value of the reaction

$$(m_A+m_a)c^2 - (m_B+m_b)c^2 = \Delta mc^2 = Q ,$$

and hence Q is positive for an exothermic reaction, and appears as increased kinetic energy of (B+b) compared with that of (A+a). For a negative Q the reaction is said to be 'endothermic'.

Masses of nuclei as such are very rarely measured - exceptions are the very lightest nuclei, e.g. protons and alphas. Usually we measure masses of ions which are readily converted to masses of atoms by adding on the masses of the requisite small number of electrons. The internationally adopted scale of atomic masses identifies the mass of the ^{12}C atom as 12 u (exactly), where the symbol u stands for 'unified mass unit', the unification referring to the different scales used previous to 1961 by physicists and by chemists. When nuclear reactions are expressed in equation form it is usual to use symbols for nuclei, and before atomic masses can be used it is important to add and balance the requisite number of electrons. (Allowance for the mass equivalence of the electron binding energies is rarely required because the small Z differences between the atoms involve very small differences in electron binding energies.)

Thus the nuclear equation

$$^{A_1}_{Z_1}A + {}^{A_2}_{Z_2}a = {}^{A_3}_{Z_3}B + {}^{A_4}_{Z_4}b + Q$$

must conserve electric charge, and hence $Z_1+Z_2 = Z_3+Z_4$. The corresponding atomic equation when A,a,B and b are all nuclei is

$$\left.\begin{matrix} A \\ +Z_1e^- \end{matrix}\right\} + \left.\begin{matrix} a \\ +Z_2e^- \end{matrix}\right\} = \left.\begin{matrix} B \\ +Z_3e^- \end{matrix}\right\} + \left.\begin{matrix} b \\ +Z_4e^- \end{matrix}\right\} + Q$$

and clearly the number of atomic electrons balances. Hence $Q/c^2 = (M_A+M_a) - (M_B+M_b)$, where M is the <u>atomic</u> mass.

However, not all the particles in the nuclear reaction need be nuclei. They can be, for example, neutrons, beta particles, or neutrinos, and the result is then not always so simple. For the naturally occurring beta emitters the emitted electrons are always negatively charged, i.e. ordinary electrons. Some artificially produced radioisotopes, however, are found to emit positively charged electrons or positrons. A few isotopes emit both β^- and β^+ particles, e.g. ${}^{64}_{29}Cu$ (β^- 0.57 MeV, β^+ 0.66 MeV, $\tau_{\frac{1}{2}}$ = 12.9 min), and the atomic mass equations are then

$$\left.\begin{matrix} {}^{64}_{29}\mathrm{Cu} \\ +29e^- \end{matrix}\right\} = \left.\begin{matrix} {}^{0}_{-1}e \\ -e^- \end{matrix}\right\} + \bar{\nu}_e + \left.\begin{matrix} {}^{64}_{30}\mathrm{Zn} \\ +30e^- \end{matrix}\right\} + Q^-$$

and

$$\left.\begin{matrix} {}^{64}_{29}\mathrm{Cu} \\ +29e^- \end{matrix}\right\} = \left.\begin{matrix} {}^{0}_{+1}e \\ +e^- \end{matrix}\right\} + \nu_e + \left.\begin{matrix} {}^{64}_{28}\mathrm{Ni} \\ +28e^- \end{matrix}\right\} + Q^+$$

Notice that the balancing of the added 'atomic' electrons leads to the cancellation of electron masses for β^- emission

$$Q^-/c^2 = M({}^{64}\mathrm{Cu}) - M({}^{64}\mathrm{Zn}) ,$$

but for the positron emission

$$Q^+/c^2 = M({}^{64}\mathrm{Cu}) - M({}^{64}\mathrm{Ni}) - 2m_e .$$

The neutrinos do not enter into consideration as they have zero rest mass (see pages 147 to 149 for the difference between ν_e and

$\bar{\nu}_e$). The Q-values therefore relate to the maximum beta-particle energy, with a small correction due to nuclear recoil.

Copper-64 happens to exhibit another type of 'radioactivity' known as 'electron capture', in which the nucleus captures one of the atomic electrons, usually from the K-shell. The result is

$$\left.\begin{matrix} {}^{64}_{29}\mathrm{Cu} \\ +29e^- \end{matrix}\right\} + \left.\begin{matrix} {}^{\;o}_{-1}e_K \\ -e^- \end{matrix}\right\} = \left.\begin{matrix} {}^{64}_{28}\mathrm{Ni} \\ +28e^- \end{matrix}\right\} + \nu_e + Q_{ec} ,$$

and therefore

$$Q_{ec}/c^2 = M({}^{64}\mathrm{Cu}) - M({}^{64}\mathrm{Ni}) .$$

The energy Q_{ec} provides the kinetic energy of the neutrino and of the recoil ^{64}Ni ion and also the energy of atomic excitation of the nickel ion, which is revealed by the emission of characteristic X-rays, associated with a K- or L-vacancy in nickel.

For negative-Q (or endothermic) reactions it is necessary to provide energy to make the reaction possible. For an endothermic reaction A(a,b)B with target A at rest, the centre of mass kinetic energy $T_{IN\ CM}$ needs to be greater than Q. Since some energy is needed to provide kinetic energy $T_{OF\ CM}$ of the centre of mass in order to conserve the linear momentum of the centre of mass, we have

$$T_{LAB} = T_{IN\ CM} + T_{OF\ CM} ,$$

and it follows that the condition for the reaction to 'go' is

$$T_{LAB} > Q\ (M_A + M_a)/M_A ,$$

since $$T_{IN\ CM} = T_{LAB}\, M_A/(M_A + M_a)\ .$$

CROSS-SECTIONS

The definition and discussion of differential cross-section $d\sigma(\theta)/d\Omega$ and of the total cross-section for a given process, i.e. elastic or inelastic scattering or nuclear reaction, are given on pages 41 and 42. For Rutherford scattering the required integration over all scattering angles suggests at first sight that the total elastic cross-section σ_{el} will be infinite. However, for θ small the distance of nearest approach to the nucleus is so large that the atomic electrons partially shield the nuclear charge and the Rutherford formula requires adjustment.

It is usual to define a total reaction cross-section σ_R at a given incident energy, to include the probability of all nuclear reactions in the restricted sense of the term together with inelastic scattering. This meaning of 'total reaction' arises in part from theoretical models, e.g. the optical model (see page 48 f), which present one expression for direct elastic scattering and a separate expression for all other processes, i.e. 'inelastic processes' consisting of inelastic scattering and reactions, and 'compound nucleus elastic scattering'.

The information about nuclei which can be obtained from inelastic processes includes:

(1) energy levels of target nuclei - from energies of scattered particles at a given angle after allowance for recoil;

(2) properties of excited nuclei, such as their angular momentum (or 'spin') and their parity (see page 161) - from angular distributions of reaction products and emitted gammas;

(3) the deformation parameter describing the departure from sphericity of the target nucleus - from

inelastic scattering differential cross-sections and also gamma-absorption reactions;

(4) the mean lifetimes of excited states - by direct observation for $\tau > \sim 10^{-6}$s, by special electronic techniques for $\tau \rightarrow 10^{-11}$s, and from indirect measurements of the spread ΔE in emitted gamma energies which relate to the mean lifetime through Heisenberg's uncertainty relation $\tau = \Delta E/\hbar$.

With such information it is possible to test theories of both nuclear structure and nuclear reaction mechanisms (see G. A. Jones in this Series).

NEUTRON REACTIONS

The zero electric charge of the neutron makes it a particularly useful particle for probing the nucleus since there is no Coulomb repulsion to overcome. However, for the self-same reason it is relatively difficult to detect its presence and hence to measure its energy, and it is particularly difficult to provide neutrons in beams of good energy resolution and small angular spread. Fermi and his co-workers in 1934 noted that neutrons slowed down by elastic collisions with nuclei in hydrogenous materials were much more reactive than 'fast' neutrons produced by alpha particles on beryllium (see page 61). Most nuclei absorbed slow neutrons to yield artificial radioactive isotopes, often producing gammas at the same time.

The simplest reaction can be expressed in the general form

$${}^{A}_{Z}X + {}^{1}_{0}n \rightarrow {}^{A+1}_{Z}X + Q_1 ,$$

where Q_1 is typically 4 - 10 MeV and the kinetic energy of the neutron ($\simeq$ 1/40 eV) is in comparison small enough to be neglected. The excess energy appears as excitation of ${}^{A+1}X$, and most commonly is emitted as gammas, hence the (n,γ) reaction is often termed 'radiative absorption'. But ${}^{A+1}X$ has excess neutrons compared with the initial ${}^{A}X$ and is usually β^--active,

$$ {}^{A+1}_{Z}X \rightarrow {}^{A+1}_{Z+1}Y + {}^{0}_{-1}e + \bar{\nu}_e + Q_2 . $$

An important example is ${}^{115}_{49}In$ (abundance 95.7%), yielding ${}^{116}In$ which is β^--active with a half-life of 54.1 min, producing stable ${}^{116}_{50}Sn$. Indium foils are used to measure integrated fluxes of low-energy neutrons by means of the beta activity developed.

Studies of neutron absorption cross-sections as a function of energy in the electronvolt range revealed rapid and large changes reminiscent of resonance phenomena. Fig. 6.1 shows the

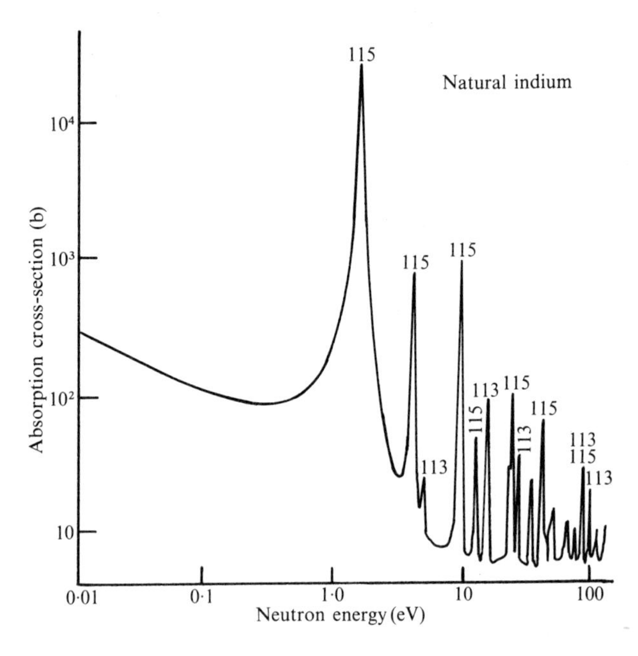

FIG. 6.1. Resonant absorption of low energy neutrons by natural indium (95.72% ${}^{115}In$, 4.28% ${}^{113}In$).

result for natural indium and illustrates the exceptionally large cross-sections (> 10^4 b) and the very small energy spreads or widths ($\Delta E < 1$ eV). These neutron resonances arise when the mass of the target nucleus plus that of the neutron plus the mass equivalent of the neutron kinetic energy is very close to the mass of the compound nucleus in specific quantized energy states. Since the energy to separate a neutron from an unexcited nucleus is 4 - 10 MeV this energy must also be the excitation energy of the compound nucleus formed by neutron absorption. At such high energy excitations the compound nucleus has energy states every few electronvolts, and very small adjustments of neutron energy are needed to pick out the levels. At higher neutron energies (> 100 eV) the peak widths become broader, and from Heisenberg's uncertainty relation $\Delta E.\Delta t \geq \hbar$ it follows that the mean lifetimes of the excited states must be shorter. The 'decay' of the excited state is either by gamma emission or by particle emission (e.g. neutron re-emission), i.e. compound elastic (or resonance) scattering. For $\Delta E = 1$ eV, $\tau \simeq 6 \times 10^{-16}$ s - rather short for gamma emission but remarkably long for neutron emission, since we expect about 10^{-20} s or less. This is because the excitation energy is shared with all the nucleons in the compound nucleus until a statistical fluctuation makes it reside on one neutron.

Other important examples of neutron resonances are shown for ^{238}U in Fig. 6.2.

FISSION AND CHAIN REACTIONS

Fermi recognized that in radiative absorption of neutrons the net result, following the expected β emission, would be a nucleus with Z increased to Z+1, and he proceeded in 1934 to investigate the element with the highest known Z, i.e. uranium (Z = 92), hoping to produce new 'transuranic' elements. The expected beta-activity was found to have several characteristic half-lives, and, since chemical tests suggested that the Z values

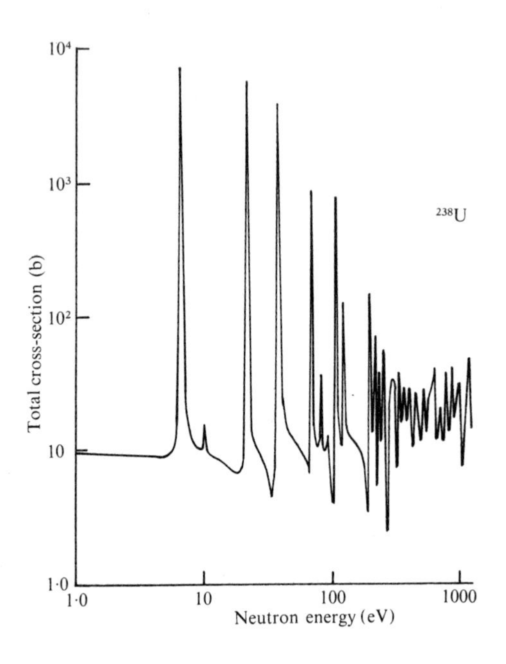

FIG. 6.2. Resonant absorption of low energy neutrons by ^{238}U. This total cross-section includes contributions from both (n,n) and (n,γ) reactions. The five largest resonances are almost entirely (n,γ), elsewhere (n,n) is dominant.

were not between 85 and 92, it was concluded that the new element 93 (and possibly 94) had been produced. Other workers made detailed chemical analyses which indicated how complex was the range of reaction products, and in late 1938 Hahn and Strassmann were forced to the conclusion that barium, lanthanum, and cerium, with nuclear charges 56, 57, and 58, were among the elements formed. Within a few days Lise Meitner and Frisch had realized the significance of earlier suggestions (e.g. Ida Noddack (1934) and Hahn and Strassmann (1939)) and wrote 'it seems possible that the uranium nucleus has only small stability of form, and may after neutron capture, divide itself into two nuclei of roughly equal size.' They pointed out that these fission fragments would have excess neutrons and be beta-active, and each fission process would release considerable energy, about 200 MeV, nearly ten times greater than the highest known energy in a non-fission reaction (22 MeV for $^{6}Li(d,\alpha)^{4}He$).

Within a few days Frisch had detected high-energy fission fragments, and very soon a number of workers confirmed the startling discoveries. Fermi suggested that neutrons might be simultaneously emitted. It was found that so-called delayed neutrons were emitted for a short time after the initiating neutron bombardment had ceased and the intensities had characteristic half-lives up to about 1 min. The delay is associated with the half-life for beta-emission, which produces a nucleus that immediately ($< 10^{-20}$ s) emits a neutron. Later it was found that on average about 2.5 'prompt' neutrons are emitted, i.e. at the moment of fission. At an early stage it was expected that this number would be greater than unity, and the possibility of using these neutrons to produce further fissions and hence a chain reaction was considered in March 1939 by several research groups, including von Halban in France and Fermi in the USA.

It was realized that if, say, two new neutrons were effective for each initial neutron then after only 80 generations of multiplication there would be some 10^{24} effective neutrons. These could release about 5×10^{6} kilowatt-hours of energy (equivalent to about 5000 tons of TNT). From 10 kg of uranium, in a time estimated at less than 10^{-6} s, the result would be a tremendous explosion. Such speculations were extended to cover the possibility of controlled chain reactions, but after the outbreak of the Second World War (in late 1939) it was deemed necessary to restrict the publication of results.

The essential information for the production of nuclear bombs and nuclear-power generation was gradually released after the War had been brought to an end by the two nuclear bombs exploded over Japan in 1945. It was confirmed that new elements

Z = 93 (neptunium, Np) and Z = 94 (plutonium, Pu) were indeed produced in the U(n,γ) reaction, as Fermi suspected. (Using a variety of nuclear reactions, all the possible 'transuranic' elements with atomic numbers from Z = 93 to 105 have now been produced, and claims exist for higher numbers. See Table 6.1.)

TABLE 6.1.

Transuranic elements (longest lived isotopes)

Neptunium	$^{237}_{93}Np$	$(2.14 \times 10^7$ y)	Fermium	$^{257}_{100}Fm$	(10.0 d)
Plutonium	$^{244}_{94}Pu$	$(7.59 \times 10^7$ y)	Mendelevium	$^{256}_{101}Md$	(1.5 h)
Americium	$^{243}_{95}Am$	$(7.63 \times 10^3$ y)	Nobelium	$^{255}_{102}No$	(15 s)
Curium	$^{247}_{96}Cm$	$(1.67 \times 10^7$ y)	Lawrencium	$^{257}_{103}Lw$	(8 s)
Berkelium	$^{247}_{97}Bk$	$(9.50 \times 10^3$ y)	Kurchatovium	$_{104}Ku$	(?)
Californium	$^{251}_{98}Cf$	$(7.9 \times 10^2$ y)	Hahnium	$_{105}Ha$	(?)
Einsteinium	$^{254}_{99}Es$	$(4.8 \times 10^2$ d)	Element-106, reported 1975: Zh.Eskp. & Teor. Fiz.Pis'ma 20, 580.		
			Element-107, reported September 1976 at Dubna, U.S.S.R.		

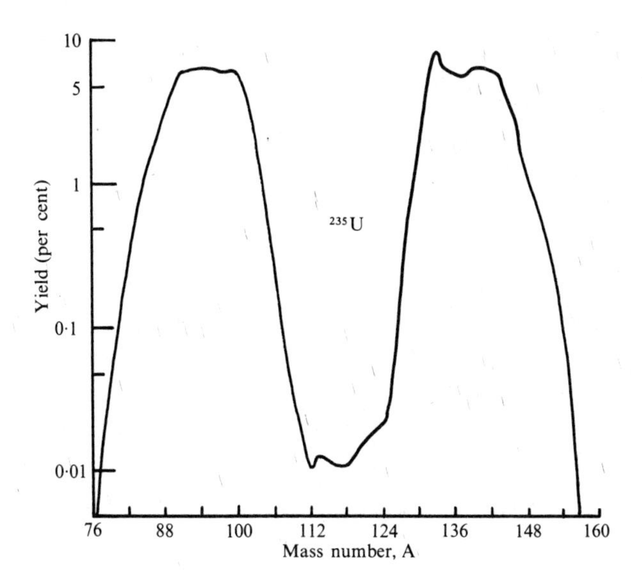

FIG 6.3. Distribution of masses of fragments from thermal neutron fission of ^{235}U.

The mass distribution of the uranium fission fragments (Fig. 6.3) revealed that the fission was predominantly asymmetric, a typical example being

$$\begin{array}{ccccccccc}
^{235}_{92}\mathrm{U} & + & ^{1}_{0}\mathrm{n} & \rightarrow & ^{140}_{56}\mathrm{Ba} & + & ^{93}_{36}\mathrm{Kr} & + & 3\left(^{1}_{0}\mathrm{n}\right), \\
 & & & & \text{12.8 days} \ \big\downarrow \ \beta^- & & \big\downarrow & & \\
 & & & & ^{140}_{57}\mathrm{La} & & \text{Five } \beta^- \text{ emissions} & & \\
 & & & & \text{40.2 hours} \ \big\downarrow \ \beta^- & & \text{(2 sec to } 10^6 \text{ years)} \ \big\downarrow & & \\
 & & & & ^{140}_{58}\mathrm{Ce}_{82} & & ^{93}_{41}\mathrm{Nb}_{52} & &
\end{array}$$

The fact that ^{140}Ce has 82 neutrons and ^{93}Nb has 52 neutrons is significant since nuclei with Z or N values near certain 'magic' numbers are known to be particularly stable - these magic numbers are 2,8,20,28,50,82, and 126 neutrons or protons (see pages 134 ff).

The energy released per fission can be estimated from the variation of nuclear binding energy with atomic mass number. The total binding energy B (see page 33) is defined by

$$B = \left\{Zm_p + (A-Z)m_n - m(A,Z)\right\}c^2 ,$$

and is therefore the energy required to break up the nucleus (A,Z) into its constituent isolated neutrons and protons. A plot of B/A, the binding energy per nucleon (the generic name for neutrons and protons), is shown in Fig. 6.4, and if we consider

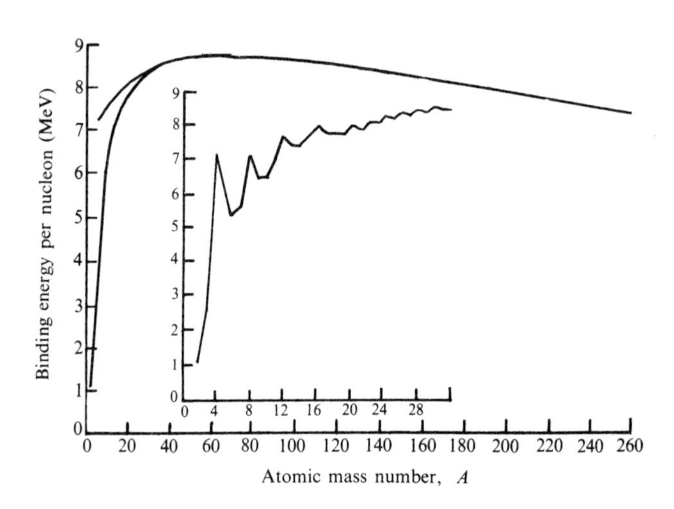

FIG. 6.4. Variation of binding energy per nucleon with mass number.

B/A values of 7.5 MeV for A = 235, 8.3 for A = 140, and 8.5 for A = 93, then the energy released is $-B_{235} + B_{140} + B_{93} = -235 \times 7.5 + 140 \times 8.3 + 93 \times 8.5 = 190$ MeV. Calculations for the above reaction using atomic masses give 200 MeV, and measurements for ^{235}U give an average value for all types of fission of (204 ± 7) MeV. About 165 MeV appears as kinetic energy of the fission fragments and the rest as the kinetic energy of neutrinos, gammas, betas, and neutrons.

Natural uranium consists of 99.27% ^{238}U, 0.72% ^{235}U, and 0.0057% ^{234}U. It is found that ^{235}U undergoes fission with low-energy 'thermal' neutrons ($T_n \simeq 1/40$eV), but ^{238}U, the common isotope, requires neutron energies of at least 1 MeV (Fig. 6.5).

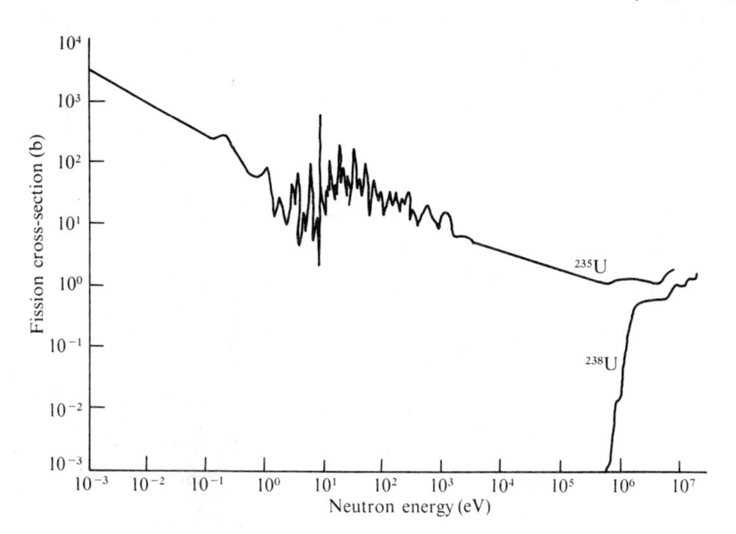

FIG. 6.5. Cross-sections for fission as a function of neutron energy for ^{235}U and ^{238}U.

The main reason is that a neutron that is paired off with another neutron is more tightly bound than an odd neutron in a comparable nucleus. Thus a thermal neutron entering ^{235}U forms a neutron pair and the total consequent excitation energy (which must be removed by gammas, say, if the resultant nucleus is to be left in its ground state) is 6.5 MeV, but for ^{238}U forming ^{239}U it is only 4.8 MeV. The separation of the fission fragments is something like alpha emission from a heavy nucleus, although the mass (and charge) ratios are rather different and the potential barrier is consequently different. The wave-mechanical potential-barrier penetration probability is sensitively dependent on the available energy.

This suggests that natural uranium is unsuitable for a nuclear bomb, since most of the neutrons emitted in fission have energies less than 1 MeV, but there is a further factor that makes it totally unacceptable, namely, the high (n,γ) cross-section of ^{238}U leading to $^{239}_{93}Np$ and $^{239}_{94}Pu$ (Fig. 6.2). For the first atomic bomb dropped in the Second World War pure ^{235}U was used, after laborious separation (see page 95); and for the second one, $^{239}_{94}Pu$. Plutonium is produced by ^{238}U (n,γ) + (β^-) and decays by alpha emission with a half-life of 24 360 years, yielding ^{235}U. The odd neutron in ^{239}Pu ensures that it undergoes fission with thermal neutrons, and, therefore, with neutrons of all energies.

The possibility of a controlled nuclear chain reaction using natural uranium was shown to depend on minimizing the loss of neutrons of energies 5 - 100 eV in ^{238}U (Fig. 6.2). The solution adopted, and still used in thermal reactors, was to form the natural uranium in relatively thin rods embedded in very pure graphite or heavy water (deuterium oxide). The elastic scattering by these light nuclei quickly slows down or 'moderates' the fast neutrons from the uranium rods with very little absorption. Ordinary water is useless because of the relatively high neutron absorption cross-section for protons, leading to deuterons and 2.2 MeV gammas. The thermal neutrons diffuse back into the fuel rods where rather less than 50% produce fission in ^{235}U. The multiplication factor k for neutron numbers per generation needs to be slightly greater than unity in order to build up a suitable neutron density, and hence power production, and the level can be controlled by the insertion or extraction of cadmium or other rods that have a very high (n,γ) cross-section for thermal neutrons. If all neutrons were prompt the build-up would be uncontrollably rapid, since a generation has a period of only ~ 1 ms, but, fortunately, about 0.4% are delayed by beta decays with half-lives greater than 0.1 s. Hence k is adjusted to be just less than unity for prompt neutrons and is able to be just greater than unity when all neutrons are included.

An important factor for both reactors and bombs is the critical size. Neutrons are lost from the surface of the device at a rate approximately proportional to the square of the linear dimensions, whereas neutron generation rates increase as the cube of the size. For natural-uranium thermal reactors the total size is of the order of a few metres cube; for bombs the uranium or plutonium needs to be of the order of 5 cm radius. Note that such a lump of ^{235}U would be spontaneously explosive, since only one initiating neutron is required and there are always a few neutrons present from spontaneous fission or from cosmic rays. In a bomb, therefore, the critical size is achieved by the use of initially well-separated subcritical masses with conventional explosives to effect rapid assembly of a single lump.

During the last thirty years several distinct types of nuclear reactor have been designed and operated for the generation of electrical power and for research purposes. Some use natural uranium enriched in ^{235}U, which allows smaller reactor sizes and the use of ordinary water as a moderator. Some are specifically designed to produce more 'fissile' material (^{233}U from ^{232}Th) than is consumed whilst still producing useful power. Other types of these thermal breeder reactors, using ^{238}U, do not quite break even. The so-called 'fast' reactors use ^{239}Pu because of its relatively large fission cross-section for 'fast' neutrons and they can therefore dispense with the moderator. This type is more efficient in neutrons and hence allows more efficient breeding by conversion of ^{238}U for ^{239}Pu. For thermal-reactor power generation, the heat produced at temperatures up to 800 K is transferred from the fuel rods to a heat exchanger by circulating fluids, e.g. carbon dioxide or helium. Ordinary water can be used in enriched reactors that are small enough to be transportable. For fast reactors liquid sodium is probably the best heat-transfer fluid. The relative safety of different types of nuclear power stations is an important and still unresolved consideration.

FUSION REACTIONS

In nuclear fission a heavy nucleus of lower B/A is converted into (two) nuclei of higher B/A and considerable energy is released (Fig. 6.4). Similarly, when two light nuclei ($A < 10$) of lower B/A combine to form a heavier nucleus of higher B/A, energy is also released, and this reaction is known as nuclear fusion. For example, in the deuterium-tritium reaction:

$$^{2}_{1}H + ^{3}_{1}H \rightarrow ^{4}_{2}He + ^{1}_{0}n + 17.6 \text{ MeV}.$$

Because of Coulomb repulsion, deuterons need more than 1 keV before the cross-section (i.e. probability) is large enough to make this reaction observable.

It is conceivable that such energies could be achieved in a high-temperature plasma, i.e. totally ionized gas, and the energy produced could compensate for energy losses and be drawn off for power generation. Such thermonuclear reactors have been investigated for over 25 years, but many problems remain, notably the problem of how to contain the hot plasma. Thermal neutrons at room temperature have a most probable energy of $kT = 1/40$ eV. The temperature corresponding to 1 keV is about 1.2×10^7 K, and the difficulties are considerable. Most devices that have been tried are based on carefully designed magnetic fields in either linear or toroidal vessels. The fields cause the moving ions to stay away from the walls, at least during pulses lasting about 1 ms, but so far a controlled and adequately sustained reaction has not been achieved. It may well be that a breakthrough in design is imminent (e.g. recent work has been done with powerful pulsed lasers and droplets of deuterium and tritium) or that present techniques on a large enough scale will be successful.

Nature provides the requisite large scale and high temperatures within stars, of which the Sun is a good average example. Although the surface temperature is only about 6000 K the Sun's internal temperature is up to 1.5×10^7 K, and densities of the predominantly hydrogen plasma are up to 130×10^3 kg m^{-3}. Under these conditions the reactions suggested by Bethe (1939) are almost certainly the dominant source of solar energy:

$$^1_1\mathrm{H} + {}^1_1\mathrm{H} \rightarrow \left({}^2_2\mathrm{He}\right) \rightarrow {}^2_1\mathrm{H} + \beta^+ + \nu_e + 0.42\ \mathrm{MeV}\ ,$$

$$^2_1\mathrm{H} + {}^1_1\mathrm{H} \rightarrow {}^3_2\mathrm{He} + 5.49\ \mathrm{MeV}\ ,$$

$$^3_2\mathrm{He} + {}^3_2\mathrm{He} \rightarrow \left({}^6_4\mathrm{Be}\right) \rightarrow {}^4_2\mathrm{He} + {}^1_1\mathrm{H} + {}^1_1\mathrm{H} + 12.86\ \mathrm{MeV}$$

The net result is

$$4\left({}^1_1\mathrm{H}\right) \rightarrow {}^4_2\mathrm{He} + 2\beta^+ + 2\nu_e + 26.7\ \mathrm{MeV}.$$

An alternative series of reactions involves carbon, nitrogen, and oxygen nuclei with the same net result, and for the Sun the two series are probably equally important.

An explosive thermonuclear or fusion device was developed in the USA, using a fission bomb to produce the high temperatures, and was first exploded in 1955. In principle, such hydrogen bombs can be made exceptionally powerful since the 'burning' is self-sustaining and the amount of fuel, which is inherently stable, is not restricted, as it is in the fission bomb, by considerations of critical size. They can also generate enormous quantities of radioactivity in the atmosphere, with consequent dangers to life on earth.

PROBLEMS

6.1. A proton at rest absorbs a 'thermal' neutron (of negligible kinetic energy) to form a deuteron and a gamma photon. Calculate the energy of the photon and the recoil energy of the deuteron. (M_H = 1.007825 u, M_n = 1.008665 u, M_D = 2.014102 u, 1 u = 931.5 MeV/c^2.)

6.2. A nucleus of mass m_1, kinetic energy T_1, is incident on a nucleus of mass m_2 initially at rest. The compound nucleus formed breaks up into nuclei of masses m_3 and m_4 which are emitted along the line of the incident beam. Derive an expression for the Q of the reaction in terms of T_1 and T_3 and the ratios m_1/m_4 and m_3/m_4.

6.3. In a non-relativistic nuclear reaction A(a,b)B the particle b is ejected at 90° to the direction of the incident particle a, with a kinetic energy T_b. Show that the Q of the reaction is given by

$$A = T_b\ (1 + M_b/M_B) - T_a\ (1 - M_a/M_A)\ .$$

6.4. (a) Make a rough estimate of the lowest energy of alpha particles that would be expected to exhibit diffraction effects in elastic scattering by ^{64}Ni (use non-relativistic formulae; mass of alpha particle = 6.68×10^{-27} kg).
(b) The threshold energy for the (α,n) reaction on ^{64}Ni is 5.20 MeV. What is the mass of the stable nuclide formed? The masses of the other atoms involved are 4.002603 u and 63.927958 u, and that of the neutron 1.008665 u. (c) The nuclide formed in the (α,p) reaction on ^{64}Ni is unstable. Without attempting numerical calculations, give reasons for the type of radioactivity that it can be expected to exhibit.

6.5. The nucleus $^{15}_{7}N$ has an excited state at 11.90 MeV. What is the energy of neutrons leading to a corresponding resonance in the reaction $^{14}_{7}N$ (n,p) $^{14}_{6}C$? (Atomic masses: ^{14}N = 14.003074 u, ^{15}N = 15.000108 u; neutron mass 1.008665 u.)

6.6. A nucleus of mass m captures a K electron and emits a neutrino. Show that the nuclear recoil is $T_m = Q^2/2mc^2$ and that almost all the available kinetic energy is associated with the neutrino.

6.7. The isobars $^{37}_{18}A$ and $^{37}_{17}C\ell$ have atomic masses 36.966772 u and 36.965898 u respectively. Does $^{37}_{18}A$ decay by electron capture or positron emission, and what is the Q-value of the reaction?

6.8. Given that only 0.1 mm of ^{113}Cd is needed to reduce a flux of thermal neutrons to 1/1000th of its original value, deduce the effective cross-section of ^{113}Cd for neutron absorption. (Density of cadmium = 8.7×10^{3} kg m^{-3}.)

6.9. Show that the mass of uranium needed to produce a megawatt-day (24 hours) of energy is approximately 1 g. (Q for fission $\simeq$ 200 MeV.)

6.10. Calculate the mass of deuterium needed to produce a megawatt-day (24 hours) of energy by (indirect) nuclear fusion to form alpha particles. ($Q \simeq 25$ MeV.)

6.11. Calculate the Coulomb repulsion energy for fission fragments from $^{235}U + n$. Assume the fragments to be spherical, with their surfaces touching and of mass ratio 1.5.

6.12. (a) Calculate the rate of change of the mass of the Sun given that its surface temperature is 5700 K and its radius 7×10^8 m. Note the necessary assumptions.
(b) Assume that the mass loss is attributable to the (indirect) fusion of protons to produce alpha particles. ($Q \simeq 25$ MeV.) What fraction of the Sun's proton content is changed to helium in (i) 10^6 years, (ii) 5×10^9 years? (Mass of Sun = 2×10^{30} kg.)

6.13. Show that the minimum gamma-ray energies required to produce (a) proton recoils of energy 5.7 MeV and (b) nitrogen nucleus recoils of energy 1.4 MeV are relatively large and significantly different. Show that these recoils can be produced, however, by neutrons of approximately the same energy and that such neutrons are available from the reaction $^9_4Be + ^4_2He = ^{12}_6C + ^1_0n$, with $T_\alpha = 5.3$ MeV ($M_{Be} = 9.012186$ u, $M_{He} = 4.002603$ u, $m_n = 1.008665$ u, 1 u = 931.5 MeV/c^2).

7. Nuclear forces and models

INTRODUCTION

The inverse-square law of force for gravity was recognized by Newton in the seventeenth century, and the similar relation for electrostatics was identified by Cavendish and by Coulomb about 100 years later. It is not uncommon when first approaching other forces, for example, Van der Waals and nuclear forces, to hope for comparably simple power laws. Unfortunately such an approach is rarely successful, and alternative methods must be adopted, many of which find expression in terms of the variation of the potential energy with the distance between the two particles. For the inverse-square laws the potentials vary quite simply as the inverse of the separation (Fig. 7.1(a)) and the application to the electronic structure of the atom, for example, is relatively straightforward.

The solutions of the Schroedinger wave equation for the ground state of the hydrogen atom are summarized in Fig. 7.1(b). Notice that the wavefunction φ is spherically symmetric in this case and can be written $U(r)/r$. The magnitude of $|\varphi|^2$ is often referred to as the probability density (or even the electron or charge density, but this can be misleading), and a careful distinction needs to be drawn between 'probability per unit volume' ρ_V (Fig. 7.1(c)) and 'probability per unit radial increment' ρ_R (Fig. 7.1(d)) as functions of radius. To emphasize this point we show in Fig. 7.2 the corresponding quantities for a sphere of uniform density, e.g. a billard ball. Notice that the ρ_R of Fig. 7.2(b) does not give the 'feeling' of a billiard ball. Nevertheless ρ_R is more common than ρ_V in representations of atoms and nuclei.

For nuclear forces no single potential has been found with

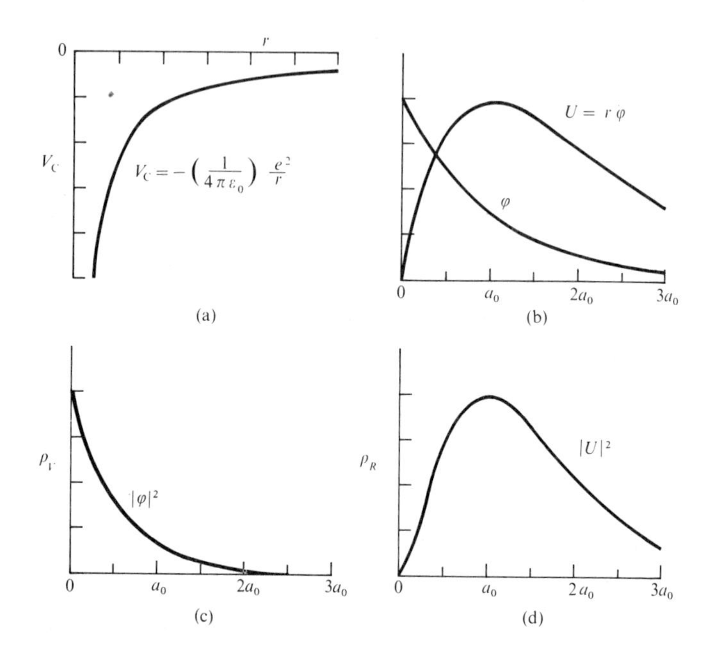

FIG. 7.1. (a) Coulomb potential energy of the electron in a hydrogen atom. (b) wavefunction $\varphi = \varphi_0 \exp(-r/a_0)$ and radial wavefunction U, for the ground state of the hydrogen atom. (c) probability per unit volume. (d) probability per unit radial increment. (These diagrams are all drawn to scale, with a_0 the Bohr radius for n = 1, $a_0 = h^2 \varepsilon_0/\pi\mu e^2$ with $\mu = m_0(M_p/(M_p + m_0))$, the reduced mass of the electron. The maximum of U is at $r = a_0$ and equals $a_0\varphi_0/e$, with e the base of natural logarithms and $\varphi_0 = (1/\pi a_0^3)^{\frac{1}{2}}$.)

the same generality as those for gravity and electrostatics, but advances in understanding have been made by assuming simple forms for the interaction potential. We shall use the simplest possible, the square well, to consider the most basic nuclear-force problem, the proton-neutron interaction, first as a bound system, i.e. the deuteron, and then unbound, as in low-energy neutron scattering by protons. Next we shall present Yukawa's important suggestions that led to the discovery of

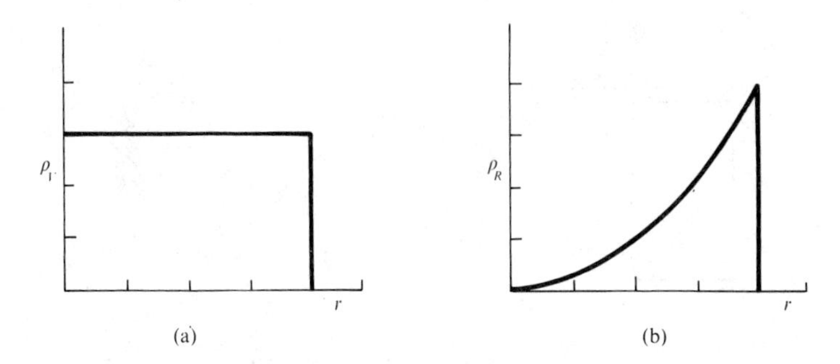

FIG. 7.2. Densities or 'probabilities' for a billiard ball: (a) volume density; (b) radial density.

mesons and to a theoretical basis for nuclear potentials. The so-called weak interaction encountered in beta emission requires different approaches, and will be treated rather briefly. Finally, we note the special difficulties presented by real nuclei, because of the many-body problem, and discuss some of the models used to interpret nuclear properties and nuclear reactions.

THE DEUTERON

In some ways the proton-neutron combination in the deuteron is the nuclear equivalent of the proton-electron in the hydrogen atom. In the atom it is a good approximation to refer to the proton as the centre of the system, because it is 1836 times heavier than the electron. But the proton and neutron have the same mass (within about 0.2%), and we need to refer the potential and the wavefunction to the centre-of-mass coordinates (r, θ, φ) and use the reduced mass $\mu \simeq m_p/2 \simeq m_n/2$ in the wave equation,

$$\frac{d^2U(r)}{dr^2} + \frac{2\mu}{\hbar^2}\{E - V(r)\}U(r) = 0$$

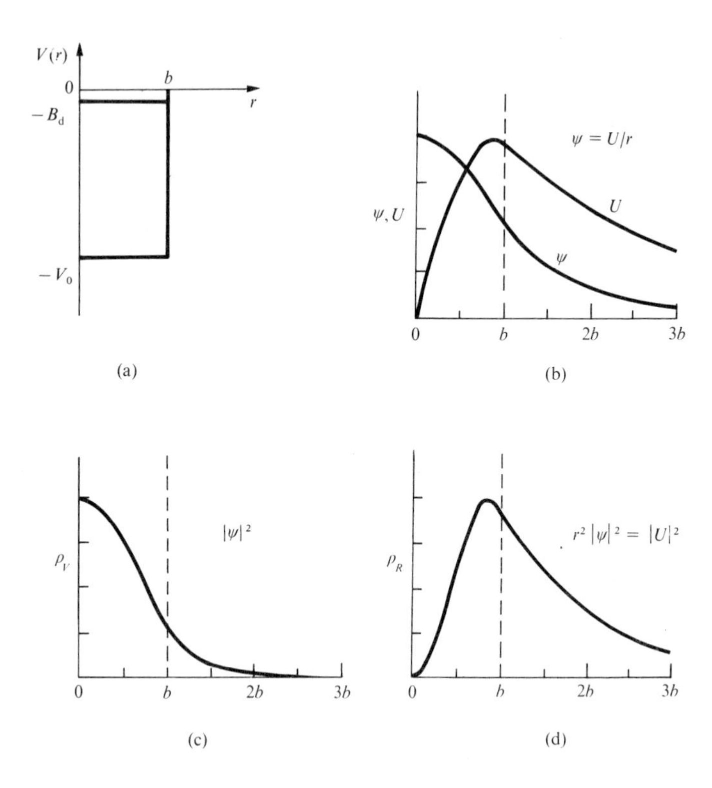

FIG. 7.3. (a) Simple square-well nuclear potential for the deuteron as a function of the relative position coordinate r, with the bound-state level of energy $-B_d$ (approximately to scale). (b) wavefunction Ψ and radial wavefunction U for the deuteron, scaled to the same maximum value. (c) probability per unit volume for the deuteron. (d) probability per unit radial increment for the deuteron.

Notice that we have already assumed for simplicity that the solution is spherically symmetric with $\Psi = U(r)/r$, and we now need to examine V(r) and E. For V(r) we shall assume the simple spherically symmetric square-well potential (Fig. 7.3(a)) and hope to find something about its depth V_0 and range b. The available relevant information about the deuteron is its binding energy $B_d = 2.23$ MeV and its root-mean-square (charge) radius, $R_{rms} \simeq 2.1$ fm, from electron scattering experiments. Clearly the energy in the ground state is $E = -B_d$, and the solution within the well radius is part of a sine wave $U(r < b) = A \sin Kr$. Outside the well we have $U(r > b) = B \exp(-\alpha r)$, and the usual boundary conditions lead to the form shown in Fig. 7.3(b) with $K \cot Kb = -\alpha$. Since $K^2 = (2\mu/\hbar^2)(V_o - B_d)$ and $\alpha^2 = 2\mu B_d/\hbar^2$ we now have a relation between B_d and the parameters of the well V_o and b, and inserting the value of B_d leads to $V_o b^2 \simeq 1.5$ MeV b ($1\ b = 10^{-28}\ m^2$). If we assume $b = R_{rms} = 2.1$ fm we obtain $V_o \simeq 34$ MeV. These values of V_o and b must, of course, yield the observed values of B_d. They also indicate, in agreement with observations, that there are no excited states of the deuteron. More detailed studies with a variety of reasonable well shapes yield similar relationships between a characteristic well-depth parameter and a characteristic range, and also exclude excited states. Other available experimental evidence, such as the deuteron spin ($J = 1$) and magnetic moment, indicates that the intrinsic spins ($s = \frac{1}{2}$) of the proton and neutron are parallel and the orbital angular momentum is zero ($\ell = 0$), confirming the reasonableness of the assumption of spherical symmetry for the wavefunction. It remains to see whether the square-well potential with the deduced parameters, $b = 2.1$ fm and $V_o \simeq 34$ MeV, can also explain low-energy neutron-proton scattering.

NEUTRON-PROTON SCATTERING

The quantum-mechanical theory of scattering, even in its simplest form, is beyond the scope of this book. A few general results, both experimental and theoretical, can be usefully presented since they lead to an important advance, conceptually simple, in our understanding of nuclear forces.

For neutron energies below a few MeV it is found that the angular distribution of the differential cross-section $d\sigma(\theta)/d\Omega$

for scattering by protons is spherically symmetric. This is characteristic of the simplest solution of the wave equation for which the (quantized) angular momentum of the neutron about the proton is zero (classically this would imply head-on collisions), and in an over-simple interpretation of quantum mechanics it implies that the product of the linear momentum and the impact parameter (Appendix A, Fig. A.1) is less than about $\hbar$. Since our deuteron potential was derived for an $\ell = 0$ state we are encouraged to see how well it predicts the magnitude of the observed total elastic cross-section ($= 4\pi d\sigma(\theta)/d\Omega$). Theory gives, in good approximation,

$$\sigma_{el}(\ell = 0) = 5.2/(T_n(CM) + B_d) \text{ barn},$$

with B_d and the neutron energy in centre-of-mass coordinates ($T_n(CM) \simeq \frac{1}{2} T_n(LAB)$) each expressed in MeV. It is clear that this indicates, since $B_d = 2.23$ MeV, that σ_{el} is closely independent of energy for $T_n \lesssim 0.1$ MeV. This independence is in excellent agreement with experiment but, unfortunately, the magnitudes do not agree:

$$\sigma_{el}(\text{theory}) \simeq 2.3 \text{ b}, \ \sigma_{el}(\text{experiment}) \simeq 20.4 \text{ b}.$$

It was Wigner in 1935 who first proposed a satisfactory solution. For the deuteron the proton and neutron spins are parallel, but for a collision between a neutron and a proton the spins can be either parallel or antiparallel. For $\ell = 0$ the total spin is then either $J = 1$ or $J = 0$. The former is referred to as a triplet state since J can take the values + 1, 0, and -1 in relation to a defined direction (e.g. the perpendicular to the scattering plane), and the latter is a singlet state ($J = 0$). In scattering the triplet state will be three times as probable as the singlet, and we can expect

$$\sigma_{el} = \tfrac{3}{4}\,\sigma_{el}(\text{triplet}) + \tfrac{1}{4}\,\sigma_{el}(\text{singlet}) \ ,$$

from which it follows that σ_{el}(singlet) needs to be 70.6 b in order to account for σ_{el}. The only way to achieve this is to adopt a singlet potential with V_o or b or both different from the triplet potential of the deuteron. With help from other experiments, e.g. the n(p,γ)d cross-section, the values deduced are:

Square well	Triplet	Singlet
Depth V_o	38.5 MeV	14.3 MeV
Range b	1.93 fm	2.50 fm

The conclusion is inevitable; the nuclear interaction is, in general, spin-dependent. The singlet state is unbound by about 50 keV (since its V_o is relatively small, with b insufficiently larger) and does not enter the deuteron problem. This spin-dependence is not to be confused with spin-orbit forces (see page 136). The latter vanish for $\ell = 0$, i.e. zero orbital angular momentum. The former are clearly still present since all our discussion has been for $\ell = 0$. Both spin-dependent forces and spin-orbit forces require a tensor term in the potential. A familiar classical system requiring a tensor force is two interacting magnetic dipoles, since the force obviously depends on their relative orientation. (See G. A. Jones 'The structure of nuclei' in this series.)

To obtain further information about the neutron-proton interaction it is necessary to study scattering at higher energies, together with more sophisticated conditions, e.g. aligning the spins of the incident particle (or the target) in polarization experiments.

YUKAWA'S FIELD QUANTUM - THE PION

The inverse-square law for electrostatics has been accounted for in the elegant but difficult and controversial theory of quantum electrodynamics. This is based on the assumption that the interaction is mediated by photons exchanged between the electric charges. The photons have energy $h\nu$, and there might seem to be, at first sight, a failure to conserve the energy of the system. However, Heisenberg's principle allows an uncertainty of energy ΔE within the time interval Δt provided that $\Delta E \Delta t \lesssim \hbar$. For an interaction at distance r, mediated at the velocity of light c, we can write $\Delta t \simeq r/c$ and since $\Delta E = h\nu$ the wavelength of the ('virtual') photon is $\lambda = 2\pi r$.

Yukawa (1935) realized that 'in the quantum theory [the nuclear] field should be accompanied by a new sort of quantum', and he was cautious enough to add, 'besides such an exchange force and the ordinary electric and magnetic forces there may be other forces between the elementary particles.' He derived an expression for the nuclear potential $V = (-g^2/r) \exp(-kr)$ (Fig. 7.4) and estimated the two constants g and k by comparison with experiment, concluding that $k = 10^{14} - 10^{15}\ m^{-1}$ and 'g a few times the elementary charge e'. It is now common to compare the so-called 'coupling constants' $g^2/hc \simeq 1$ for 'strong' nuclear and $(1/4\pi\varepsilon_0)(e^2/\hbar c) \simeq 1/137$ for electromagnetic forces.

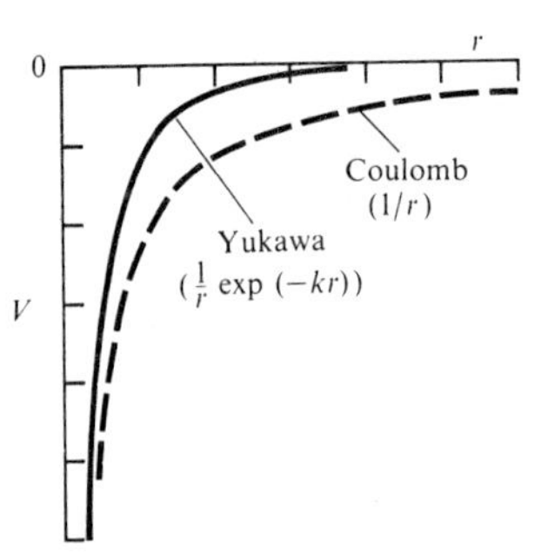

FIG. 7.4. Comparison of the form of the Yukawa 'one-pion exchange' potential with the form of the Coulomb potential.

Yukawa considered the 'nuclear field quantum' to have an electric charge +e or -e, and deduced that the rest mass M of this particle must be about 200 times the electron mass m_e. This followed from the fact that nucleon-nucleon forces act strongly over a short range, $R = \sim 2$ fm, i.e. about the radius of the deuteron. Thus we can repeat the consideration of Heisenberg's uncertainty relation with $\Delta E = Mc^2$ and $\Delta t \simeq R/c$, yielding $M \simeq \hbar/Rc$. For R = 2 fm, $M \simeq 200m_e$. With a humble over-confidence in the pre-eminence of experimentalists Yukawa remarked, 'as such a quantum with large mass and positive or negative charge has never been found by the experiment, the above theory seems to be on a wrong line.'

Two years later, with the discovery of the muon possessing just these characteristics ($m_\mu \simeq 207m_e$, $q = \pm e$), Anderson and Neddermeyer (1937) seemed to have removed Yukawa's doubts. Unfortunately the muon did not react at all strongly with protons or other nuclei, and it was not until 1947 when Powell observed the pi-mesons (or pions) in nuclear emulsions exposed to cosmic rays from outer space that the identity of Yukawa's particle became established ($m_\pi \simeq 273m_e$, $q = \pm e$ or 0). Powell's meson decays to form a muon, which in turn decays to form an electron. Perkins (1947) first observed the 'explosive' interaction of a pion with a nucleus, and within a few years many new unstable particles were observed (see Chapter 8).

The simple 'one-pion exchange potential' of Yukawa is useful as a guide, especially for nuclear forces beyond about 2 fm, but has a number of unsatisfactory features. For example, it is now generally agreed that some kind of repulsive core potential is required at a range below about 0.5 fm in order to explain high-energy nucleon-nucleon scattering, and such a core helps to explain the 'saturation' of nuclear forces revealed by the approximately constant density of nuclear matter. Early meson theories with certain types of exchange force produced the repulsive terms required to prevent the collapse of nuclear matter to nuclear dimensions, but they could not explain the high-energy scattering. Particles other than pions (e.g. kaons, see page 152) are also known to interact strongly with nucleons, and to explain polarization effects during nuclear scattering vector mesons with non-zero spin are required, yet pions and kaons have zero spin. It may be that the full nucleon-nucleon interaction is a problem of complexity rather than of principle.

THE WEAK INTERACTION - BETA EMISSION

One important type of 'nuclear' interaction not covered by the forces discussed so far is the emission of betas and neutrinos by nuclei (see pages 21, 22), of which the simplest example is the beta decay of the free neutron

$$n \rightarrow p + e^- + \bar{\nu}_e + Q \; (\tau_{\frac{1}{2}} = 10.8 \text{ min}, \; Q = 0.78 \text{ MeV}) \; .$$

The probability of these beta decays, i.e. the reciprocal of the mean life τ_{mean} is determined by the interaction potential involved. The strong nuclear potential would produce beta-decay lifetimes that are much too short, indicating that it is much too strong. In 1934, Fermi developed a theory of beta emission that successfully accounted for the continuous energy spectrum of the observed beta particles (see G. A. Jones in this Series). The theory also yielded expressions for the half-life in terms of the maximum energy of the betas and other parameters. One important parameter was the strength or 'coupling constant' of this so-called weak (nuclear) interaction which turned out to be about 10^{-13} compared with about unity for the strong interaction. Yukawa, in 1938, suggested there should be a field quantum for the weak interaction, corresponding to the photon and the pion. This so-called 'intermediate vector boson' or 'W-particle' has not been detected as a free particle. Other examples of the weak interaction occur in the decays of the muon (see page 148), of the pion and other mesons, and of hyperons.

SUMMARY OF TYPES OF INTERACTIONS

In Table 7.1 the four well-established types of interactions are summarized. Recently two further types have been proposed, the so-called super-strong and the super-weak, and relations between the different types are being actively investigated.

TABLE 7.1

Interaction	Source	Field quantum			Strength or coupling constant
		Particle	Rest mass	Spin	
Strong nuclear	BARYONS Nucleons, hyperons, baryon resonances (HADRONS)	MESONS Pion, kaon, etc., meson resonances (HADRONS)	Finite, $\simeq M_p/7$ to $\simeq 2.5M_p$	0, 1, 2, (3)	$\simeq 1$
Electro-magnetic	Electric charges	Photon	0	1	$\simeq 10^{-2}$
Weak nuclear	LEPTONS	(Intermediate vector bosons, W & Z^o particles)	(Finite)	(Integral, not zero)	$\simeq 10^{-13}$
Gravity	Masses	(Graviton)	(0)	(2)	$\simeq 10^{-38}$

Bracketed entries have been postulated or deduced from theory, but have not yet been identified.

We note that the electromagnetic interaction takes place between bodies with electric charge which act as 'sources' of the electromagnetic field quantum, the photon, with zero rest mass and spin 1. Gravity acts between bodies with mass ('source particles'), and the hitherto undetected 'graviton' has been postulated as the field quantum to mediate the interaction, with zero rest mass, in accord with the inverse-square law. Theory beyond the level of this treatment suggests a graviton spin of 2. The strong nuclear interaction takes place between 'source' particles collectively known as 'baryons' (meaning 'heavy ones'), which include neutrons and protons and other particles sometimes regarded as excited nucleons, namely, 'hyperons' (meaning the mass is 'greater than' the nucleon mass) and 'baryon resonances' (Chapter 8). The field quanta are 'mesons' (meaning mass is 'between' those of the electron and proton), e.g. pions, kaons, which also appear to combine to form more complex unstable particles ('meson resonances', see Chapter 8). The particles which interact with each other through the strong nuclear interaction, i.e. baryons and mesons, are known as 'hadrons' (meaning 'strong ones'). The sources of the weak interaction are leptons, or more accurately 'electron leptons' for $e^{\pm}$, ν_e, $\bar{\nu}_e$ and 'muon leptons' for $\mu^{\pm}$, ν_{μ}, $\bar{\nu}_{\mu}$ (the smallest Greek coin, the widow's mite of the New Testament, was a lepton, meaning 'the small one').

The strengths of the interactions are given in terms of the coupling constants, and the relative weakness of the gravitational force indicates why gravitons are so difficult to detect. Note that the electromagnetic interaction is about 1/100th of the strong nuclear, but is very much stronger than the weak interaction.

NUCLEAR MODELS

The application of quantum electrodynamics to the hydrogen

atom has been entirely successful within the very narrow limits set by precision measurements of the energy levels and relative intensities of spectral lines (i.e. transition probabilities). Atoms containing many electrons have a dominant centre of force, the nucleus, but residual interactions between the electrons introduce problems that can be solved only approximately. For nuclei there is no obvious centre of force. The many nucleons interact with all their neighbours and maybe with nucleons at a slightly longer range; to make matters worse, the exact form of the nucleon-nucleon interaction is not known. In such circumstances the physicist first resorts to models. Brief mention has been made of the optical model (see pages 48 and 49), which accounts for the scattering and absorption of incident particles in terms of a 'cloudy crystal ball'. In practice this means inserting into the wave equation an equivalent potential with real and imaginary parts (the latter to simulate absorption) and a special spin-orbit term to yield polarization effects.

The shell model

We have already considered the bound and unbound states of the neutron-proton system in terms of a simple square-well potential. Surprisingly, there was no significant success in the attempt to consider the properties of nucleons in larger nuclei in terms of suitable potentials until 1949. The possibility of nucleon 'orbits' within the dense nuclear matter was considered to be negligible. However, from accumulated evidence it became abundantly clear that nucleons occupy well-defined energy states and certain numbers of nucleons lead to extra-stable nuclei. This is reminiscent of electrons in atomic shells where the shell 'closures' are at 2, 8, 18, 32 electrons, although the higher numbers are obscured by the next shell beginning to be populated before the previous shell is completely filled, depending on which has the lower total energy for the next electron. The

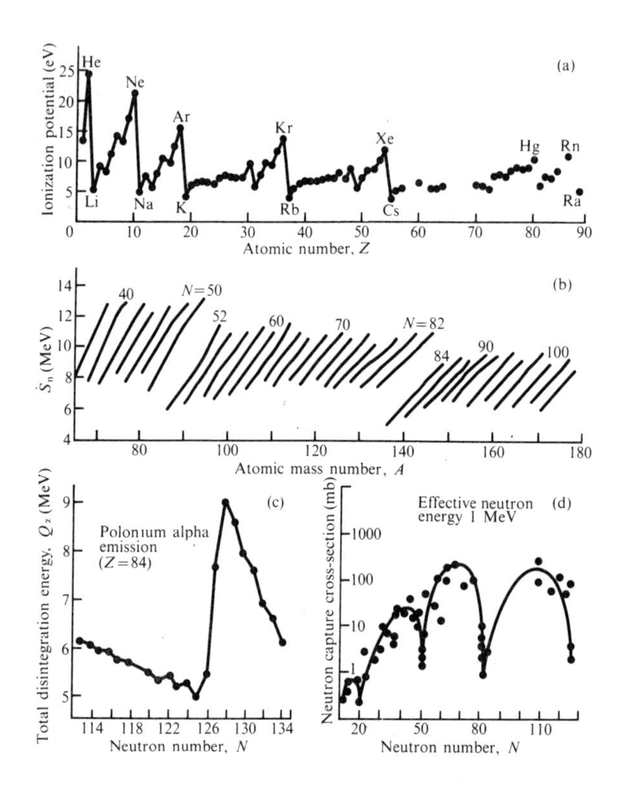

FIG. 7.5. Evidence for 'magic' numbers. (a) First ionization potentials (electron separation energies) of atoms indicating filled electron shells at Z = 2, 10, 18, 36, 54. (b) Neutron separation energies for even-N nuclei as function of A (odd and even) showing drops at N = 50, 82 (cf. (a)). (c) Disintegration energies for alpha emission from polonium isotopes showing relatively high Q_α for N = 128, indicating ease of removal of two neutrons outside N = 126 'core'. (d) Neutron capture cross-sections for fission neutrons of effective energy 1 MeV showing very low values (note logarithmic scale) for N = 20, 50, 82, 126.

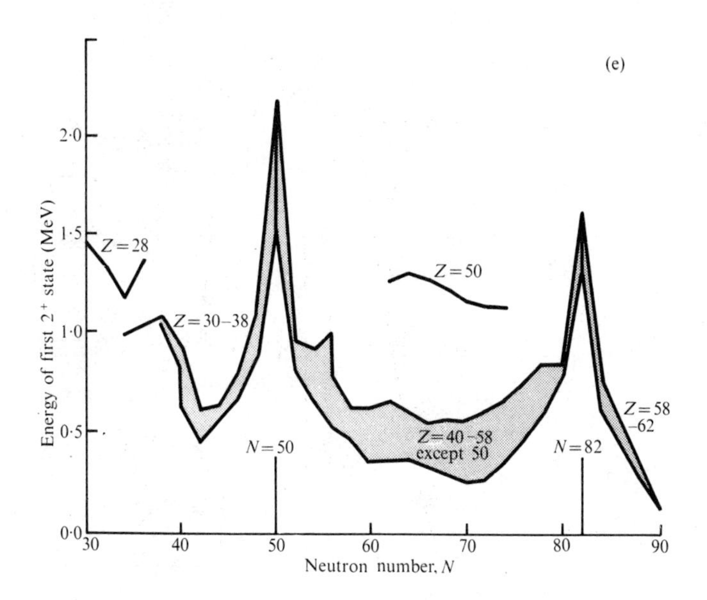

FIG. 7.5. (e) Energies of the first 2^+ states of even-even nuclei showing high values for magic numbers of protons (28, 50) and of neutrons (50, 82).

corresponding numbers for either neutrons or protons are 2, 8, 20, 28, 50, 82, and 126 and are known as 'magic numbers'. In Fig. 7.5 are shown a few examples of marked changes in nuclear properties at these numbers, and include for comparison one example of periodic properties in atoms (see also Fig. 7.7).

A simple harmonic oscillator (parabolic) potential yields energy levels (Fig. 7.6(a)) corresponding to 2, 8, 20, 40, 70, 112, etc. neutrons or protons - encouraging but not good enough. Other reasonable shapes split some of these levels to produce for example, 2, 8, 18, 20, 34, 40, and 58 (Fig. 7.6(b)), but none could produce 28 and 50 until Mayer, and independently Haxel, Jensen, and Suess (1949), introduced a term in the potential that involved the spin of the particles. More accurately, it related the particle spin ($\underline{\sigma}$) to its orbital angular momentum (ℓ), and by

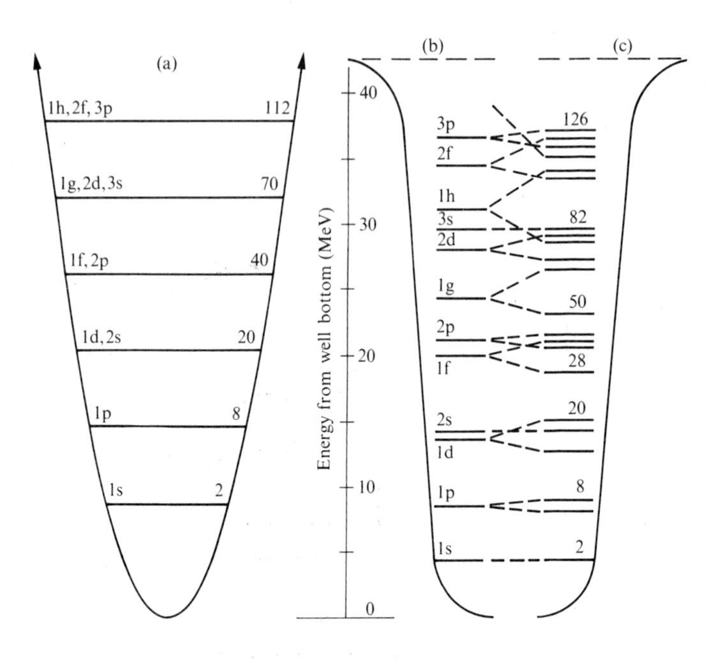

FIG. 7.6. (a) Nuclear energy levels for an harmonic-oscillator potential showing the degeneracy (i.e. equal energies) of some levels (1d, 2s), (1f, 2p), etc. The nomenclature s, p, d, f, g, h corresponds to ℓ = 0, 1, 2, 3, 4, 5, and the numbers of particles in each state is given by $2(2\ell + 1)$. (b) Energy levels for a potential of the shape shown with R ~ 10 fm and no spin-orbit term. The degeneracy of (a) has been removed and significant energy gaps occur at proton (or neutron) numbers 2, 8, 20, 34(1f), 40, 58(1g), 68(2d), 70(3s), 92(1h), 106(2f), 112(3p). (c) Effect of spin-orbit term on levels of (b). Apart from the s-levels, each level is split into two, the lower with $j = \ell + \frac{1}{2}$, the upper with $j = \ell - \frac{1}{2}$. The magnitude of the splitting increases with ℓ and the magic numbers can be obtained as shown, e.g. 28 is produced by the $1f_{\frac{7}{2}}$ state (with $j = \frac{7}{2}$) with 8 particles beyond the 20 in the lower shells.

suitably adjusting the strength V_s in the 'spin-orbit' term $V_s f(r)\underline{\sigma}.\ell$ it was possible to split the levels even further and also produce energy gaps in the sequence of levels that corresponded exactly to the magic numbers (Fig. 7.6(c)). (A similar spin-orbit term in atomic structure theory is much weaker and produces very small splitting of atomic levels.) This nuclear 'shell model' also accounted for many, but not all, of the spins of the ground states of nuclei, but was much less successful for magnetic moments. It failed almost completely to explain the numerous excited states of nuclei, and these and other properties were soon recognized to require a departure from the simplifying assumption that nuclei are spherical.

The collective model

Aage Bohr and Mottelson (1953) investigated the consequences of collective motion of nucleons which produces ellipsoidal oblate and prolate nuclei. In particular, they studied the energy levels of such nuclei when they rotate and vibrate. The results are particularly simple for pure (quadrupole) vibration, yielding equally spaced levels,

$$E_{vib} = (n + 5/2)\hbar\omega \ ,$$

where n = 0, 1, 2, etc. and ω is a characteristic frequency. For pure rotation the relation is

$$E_{rot} = R(R + 1)\hbar^2/2\mathcal{I}_{eff} \ ,$$

where R = 0, 2, 4, etc. and $\mathcal{I}_{eff}$ is the effective moment of inertia (less than that for a rigid body). A test of these relationships is provided by taking the ratio of the energies of the excited states to that of the first state. We expect 3.33, 7, 12, and 18.33 for rotational states and find, for example, for $^{164}_{68}Er$ (Erbium), 3.27, 6.7, 11.2, and 16.0. In Fig. 7.7 we show the ratio for the first two excited states for many even-Z - even-N (even-even) nuclei and the value 3.33 expected for 'pure' rotational states is clearly shown. For the equally spaced

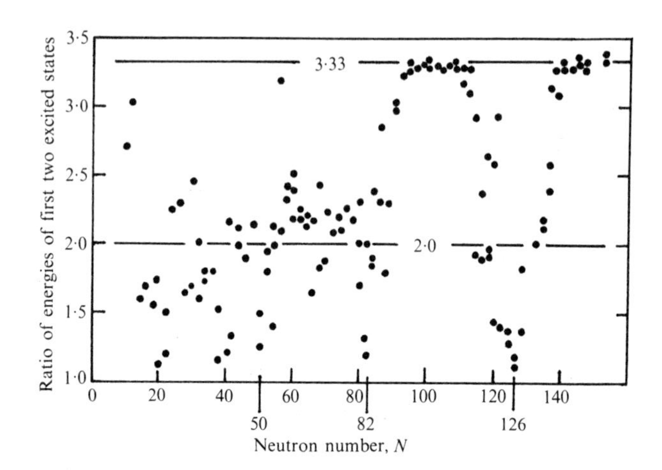

FIG. 7.7. Evidence for rotational energy levels in the non-spherical even-even nuclei with neutron numbers away from magic numbers.

levels of vibrational states this ratio is expected to be 2, but this is not so clearly or frequently exhibited, since in general rotational and vibrational motion is mixed or 'coupled' and the results can be rather complicated.

The purer vibrational states occur for near-spherical nuclei, found near magic numbers of Z or N, and are widely spaced ($\lesssim$ 1 MeV). Rotational states are more evident for distorted nuclei that occur away from magic numbers, and usually have spacings nearer to $\sim$ 100 keV.

The liquid-drop model and the semi-empirical mass formula

As early as 1935, von Weizsaecker developed a formula for the masses of nuclei, and hence their binding energies, based on the observation that the density of nuclear matter is essentially constant (see pages 49 and 50), i.e. rather like a liquid (a model probably first suggested by Gamow (1930)). This

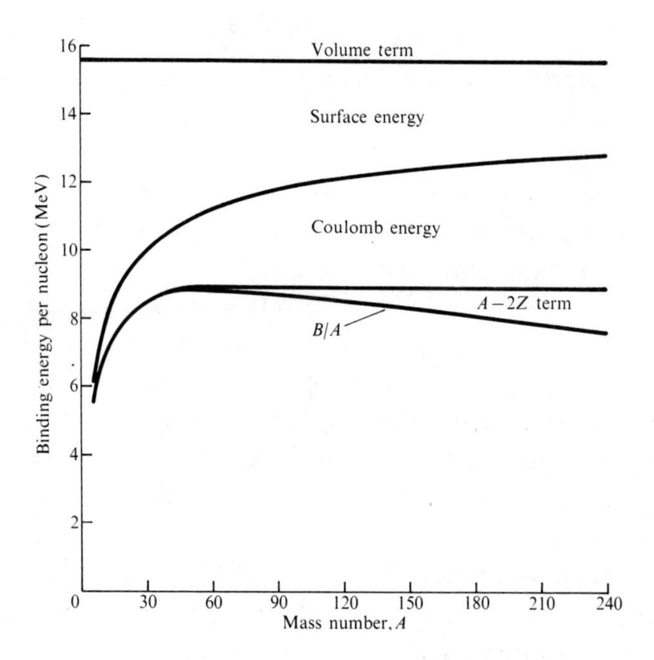

FIG. 7.8. Importance of the four principal terms in the semi-empirical mass formula illustrated by the binding energy per nucleon, $B/A = a_1 - a_2A^{-\frac{1}{3}} - a_3Z^2A^{-\frac{4}{3}} - a_4(A - 2Z)^2A^{-2}$ with $a_1 = 15.6$, $a_2 = 17.2$, $a_3 = 0.70$, and $a_4 = 23.3$ MeV. Notice the remarkable (but fortuitious) constancy of B/A for the sum of the first three terms, for $A > 45$, and the relatively small effect of the neutron-excess term $A - 2Z$.

'liquid-drop model' suggests binding energies expressible in terms of 'latent heat of vaporization' or volume energy and 'surface energy', together with allowances for the Coulomb repulsion of the protons. The dominant terms are

$$B = m_pZ + m_nN - m(A, Z) = a_1 A - a_2A^{\frac{2}{3}} - a_3Z^2A^{-\frac{1}{3}},$$

binding energy	proton masses	neutron masses	mass of nucleus	volume energy	surface energy	Coulomb repulsion

where a_1, a_2, and a_3 are constants, with a_1 corresponding to the binding energy per nucleon for infinite uncharged nuclear matter, and a_2, a_3 reducing the binding for finite nuclei. Another term allows for a diminution of binding related to A - 2Z, the excess of N over Z in most nuclei, and yet another for the fact that, in nuclei, pairs of protons (or neutrons) are in a state of lower energy when their spins are antiparallel. With only five adjustable constants, the agreement with the measured binding energies of hundreds of nuclei is remarkably good (Fig. 7.8, cf. Fig. 6.4).

The liquid-drop model has been applied to explain aspects of nuclear reaction mechanisms, notably by Niels Bohr (1935) in his idea of the compound nucleus (see pages 61, 107) and by Frisch and Meitner in their recognition of the possibilities and consequences of nuclear fission. The changes of surface energy and Coulomb energy during deformation can be shown to lead to fission provided Z^2/A is sufficiently large, as for the heaviest nuclei, and the sequence of changes of shape through elongation, necking, and break-up are not difficult to imagine.

Direct-interaction models

Many low-energy reactions involve compound-nucleus formation, but at higher energies the mechanism appears to be more direct and is completed in times of about 10^{-21} - 10^{-23}s compared with the 10^{-17}s or longer of the compound nucleus process. These direct reactions include 'knock-out' processes (e.g. p in, and p and n out, or 2p out), 'pick-up' processes (e.g. p in, and d out), and 'stripping' reactions (e.g. d incident, n goes into nucleus, and p continues out). From the differential cross-sections much information on nuclear energy levels can be deduced, and severe tests are imposed on the detailed theoretical models adopted.

PROBLEMS

7.1. The Schroedinger equation for a particle of mass μ in a central potential $V(r)$ is $\nabla_r^2\psi(r) + (2\mu/\hbar^2)\{E - V(r)\}\psi(r) = 0$ and $\nabla_r^2\psi = \frac{1}{r^2}\frac{\partial}{\partial r}\left(r^2\frac{\partial\psi}{\partial r}\right) + \frac{1}{r^2\sin\varphi}\frac{\partial}{\partial\varphi}\left(\sin\varphi\frac{\partial\psi}{\partial\varphi}\right) + \frac{1}{r^2\sin\varphi}\frac{\partial^2\psi}{\partial\theta^2}$.

Show that a spherically symmetric function $\psi = U(r)/r$ allows the wave equation to reduce to the simple form

$$\frac{d^2U(r)}{dr^2} + \frac{2\mu}{\hbar^2}\{E - V(r)\}\,U(r) = 0\ ,$$

as used for the deuteron.

7.2. Show that the matching of the magnitudes U and slopes dU/dr of the wavefunctions for the deuteron $U\ (r < b)$ and $U(r > b)$ at $r = b$ leads to $K \cot Kb = -\alpha$ (see page 125). Hence show that the solution is near to $Kb = \pi/2$ and that this value for Kb leads to $V_0 b^2 \simeq \frac{\pi^2 h^2}{8\mu} = 1.0$ MeV b.

7.3. The deuteron continuity relation can be written $\tan Kb = -Kb/\alpha b$. Show that $\alpha b = 0.487$ for $B_d = 2.23$ MeV and $b = 2.1$ fm. Solve for Kb by plotting $\tan Kb$ and $-Kb/0.487$, and hence deduce the corresponding values of V_0 and $V_0 b^2$.

7.4. Calculate the impact parameter for 14 MeV neutrons on protons such that the total orbital angular momentum is $\hbar$. Compare this with the ranges of the square-well potentials used to account for neutron-proton scattering, and comment on the likelihood of 14 MeV neutron-proton scattering being spherically symmetric.

7.5. The classical formula for the Coulomb energy of a nucleus is $(1/4\pi\varepsilon_0)\frac{3}{5}Z(Z-1)e^2/R$. The optimum value of the coefficient of the corresponding term $a_3Z^2A^{-\frac{1}{3}}$ in the semi-empirical mass formula is found to be 0.70 MeV/c^2. Compare this value with the theoretical expression above, using $R = 1.2\ A^{\frac{1}{3}}$fm.

7.6. Calculate the total binding energy B and the binding energy per nucleon B/A for ${}^{208}_{82}\mathrm{Pb}$. Express B in terms of the rest-mass energy of a proton. What fraction of B is surface energy? (See caption to Fig. 7.8. Use the atomic masses for ${}^{1}\mathrm{H}$, 1.007825 u, and ${}^{208}\mathrm{Pb}$, 207.976650 u, and the neutron mass, 1.008665 u.)

7.7. Use the semi-empirical mass formula (see page 140 and Fig. 7.8) to estimate the expected total disintegration energy in alpha emission for ${}^{197}_{84}\mathrm{Po}$, ${}^{204}_{84}\mathrm{Po}$, ${}^{209}_{84}\mathrm{Po}$, ${}^{213}_{84}\mathrm{Po}$, and compare with the values in Fig. 7.5(c). The binding energy of the alpha particle is 28.3 MeV.

7.8. Derive an expression for the atomic number of the most stable member of a set of isobars, using the semi-empirical mass formula including the neutron-excess term (Fig. 7.8) but not the pairing-energy term. Deduce the most stable members for A = 17, A = 41, and A = 79, and comment on the neglected factors that affect stability ($m_n - m_p = 1.3$ MeV/c^2).

8. Cosmic rays and elementary particles

INTRODUCTION

The discovery of radioactivity prompted the development of sensitive instruments to measure the ionization produced in air by the alpha, beta, and gamma rays. These studies indicated that ionization could not be altogether removed by shielding or by careful purification of the gas and other materials. The same instruments were installed in balloons, and registered greatly increased ionization at altitudes up to about 16 000 m. It was concluded (by Hess in 1913) that the effect was due to ionizing radiation from outside the Earth, and Millikan (1925) suggested the name 'cosmic rays'.

COSMIC RAYS

At first it was commonly assumed that the cosmic rays were gamma rays, but sea-level studies (during a cruise from Europe to Java) indicated a systematic decrease with decreasing latitude, which was most readily interpreted as due to the Earth's magnetic field acting on positively charged particles of energies up to a few GeV, the higher-energy particles being relatively unaffected.

A great deal of effort has been expended in determinations of the distribution of the particles in space and time, their nature, and their energy spectra. In general the primary rays, outside the Earth's atmosphere, are nuclei with abundances very close to those of the elements in the universe, but with markedly more lithium, beryllium, and boron. Some 94% are protons, and 5% are ^{4}He and ^{3}He. It is probable that the lithium, beryllium, and boron nuclei result from proton disintegrations of intergalactic matter. Small percentages of electrons and gamma rays are also reported. The number with energy E in the interval dE is proportional to $E^{-\gamma}dE$, with $\gamma \simeq 1.5 - 2.0$ in the range $E = 1 - 10^{10}$ GeV and is about $10^4\ m^{-2}s^{-1}\ sr^{-1}$ at E = 1 GeV, for dE = 1 MeV. Any theory of the origin of the cosmic rays must explain this energy distribution, together with the fact that their intensity is remarkably constant and isotropic, i.e. independent of time and of the direction in space. These are the dominant features of cosmic rays, but there are also fluctuations which can be explained in terms of rays from the Sun, and possibly, in the case of gamma rays, by individual stellar locations.

We shall not pursue these interesting astrophysical and cosmological topics but confine ourselves to those aspects that advanced our understanding of nuclear physics. They arose almost entirely from the interaction of the high-energy primary cosmic-ray protons with nuclei in the atmosphere. A wide variety of particles is produced, including the pions (and kaons) already discussed, their decay products (muons and neutrinos), and high-energy gammas that produce positron-electron pairs.

POSITRONS AND ANTIPARTICLES

The introduction of the positron is part of a story with a moral. A relativistic generalization of the wave equation suitable for charged particles in an electromagnetic field was first developed by Schroedinger (although it is now known as the Klein-Gordon equation), and was almost immediately rejected because it led to negative as well as positive total energies for the particles,

$$E = \pm (p^2c^2 + m_o^2c^4)^{\frac{1}{2}} ,$$

where m_o is the rest mass and p the momentum. In 1930 Dirac developed a more complete equation which not only yielded the same result for the energy states but also automatically

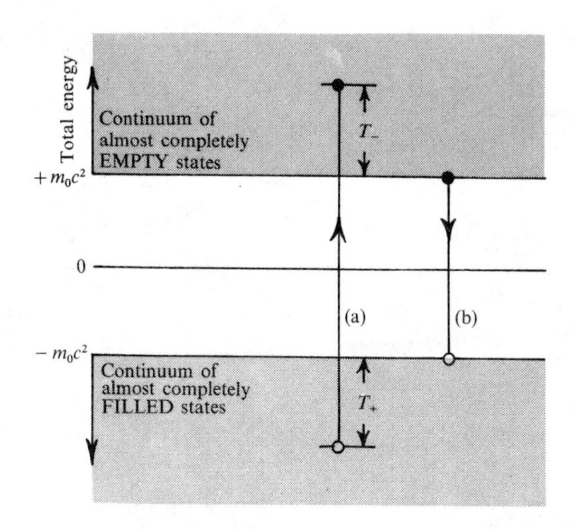

FIG 8.1. Energy states for free electrons. The transition (a) corresponds to pair production of a particle (e.g. an electron) of kinetic energy T_- and of an antiparticle (e.g. a positron) of kinetic energy T_+. The transition (b) corresponds to positronium annihilation.

predicted an intrinsic angular momentum or spin of $\frac{1}{2}\hbar$ for the particle and the correct magnetic moment (the Klein-Gordon equation appears to be valid for particles with zero intrinsic spin.) The possible energy states are shown schematically in Fig. 8.1, and the positive ones are precisely those expected for ordinary particles, e.g. electrons, with kinetic energy $T = E - m_oc^2$.

The negative energy states suggest that transitions could occur from the positive energies with the emission of electromagnetic energy, i.e. photons. But such transitions had not been observed, therefore all the negative states were considered by Dirac to be already occupied. This suggests, however, that the supplying of sufficiently energetic photons to a suitable system would promote an electron from a negative energy state to a positive one, thereby making it observable as an ordinary electron. The unfilled negative energy state, or 'electron hole', would then have the properties of a 'positively charged electron'. At the time (1930) such a particle was not known and Dirac thought it might be the proton, in spite of the discrepancy in rest mass.

Two years later, Anderson observed an electron-like track in a cloud-chamber photograph (Fig. 8.2(a)), with clear evidence that the particle was positively charged. Its direction of motion was given by its smaller radius of curvature (produced by a magnetic field) after passing through a metal plate, and its direction of curvature then indicated the positive charge. The

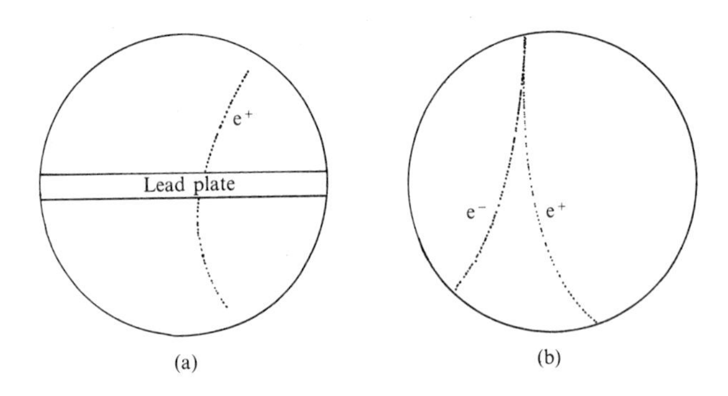

FIG. 8.2. Cloud-chamber pictures of positrons. (a) From the photograph that led to Anderson's discovery of the positron. (b) An example of electron-positron pair production in the wall of the cloud chamber (the applied magnetic fields were perpendicular to the plane of the diagram).

identification of the 'positron' of Dirac's theory was obvious but not yet complete. Electron-positron pair production by gammas of sufficient energy ($h\nu > 2\,m_e c^2 = 1.022$ MeV) was soon observed in cloud chambers (Fig. 8.2(b)), the interaction having taken place between the photon and the very strong electric field near a nucleus. Finally the 'annihilation' of a positron by combination, at rest, with an electron was observed from the two photons, each of characteristic energy 0.511 MeV, which accounted for the rest-mass energy of the two particles. It has since been found that the electron-positron system forms a kind of 'electron atom' known as positronium, which with antiparallel spins has a mean life of about 8 ns. There is also a triplet state with parallel spins and zero orbital angular momentum and a mean life of about 7 μs, forming three photons with total energy 1.022 MeV.

There is nothing in Dirac's approach to the relativistic wave equation that restricts the results to electrons. Other particles, such as the proton and neutron, can be expected to have negative energy states, but it would require gammas of energy greater than 2 × 938 MeV to produce a proton pair, or 'proton-antiproton' as they are called. Such high energies are now readily available by means of accelerators, and it is even possible to produce beams of antiprotons, admittedly of low intensity, for studying their properties in scattering and reactions, including 'proton-antiproton annihilation' which usually leads to the production of pions and occasionally kaons. The antiproton is negatively charged, and was first identified in 1955 from proton bombardment of copper nuclei at energies above 4.3 GeV. (For proton-proton collisions the threshold is $T_p = 5.6$ GeV, but protons in the copper nuclei can be moving towards the incident protons. A mere 30 MeV of target proton energy can reduce the threshold to about 4.3 GeV.)

The antiparticle of the neutron has also been observed. It has zero charge and, as for all antiparticles, its mass and spin are identical with those of the corresponding particle, but the magnetic moment of the antineutron is of opposite sign to that of the neutron in relation to the direction of their spins.

Mesons also have their antiparticles with equal masses and (integral) spins but opposite electric charges and magnetic moments. They are not necessarily created or annihilated in pairs.

Particles that have no electromagnetic interaction sometimes have antiparticles, e.g. the neutrinos (ν_e and ν_μ), and sometimes they are 'self-conjugate', i.e. particle and antiparticle are experimentally indistinguishable, as for the neutral pion (π^0).

MUONS AND LEPTON NUMBERS

The discovery in 1937 of the muon of mass about 200 times that of the electron appeared at first to support Yukawa's prediction of a nuclear field 'quantum' (see page 128). The particle was observed as a curved track originating in a lead plate in a cloud chamber operated on a high mountain (Pike's Peak in Colorado), and its droplet density and curvature (i.e. magnetic rigidity, $BR = mv/e$) together indicated a mass between that of an electron and a proton. Anderson and Neddermeyer proposed the name mesotron, which was later abbreviated to meson (Greek: meson = thing in the middle); later still, this was sharpened to mu-meson when the pi-meson was discovered, and yet later abbreviated to muon to prevent confusion when the term meson became restricted to the field quanta (pions, kaons, etc.) of the strong nuclear interaction. The fact that the muon did not interact strongly with nuclei ruled out the possibility of its being Yukawa's particle - the pion was to fill that role. It was later found that the muon decayed, with a mean life of 2.2 μs, to an electron or positron, depending on the muon charge; and the continuous energy spectrum of the electrons (similar to that in beta emission) indicated the presence of (at least) two other particles. Careful mass and energy measurements supported the possibility of two neutrinos, and in the early 1970s it became increasingly evident that these are different types of neutrino. One type, the electron neutrino ν_e, can interact with nuclei to produce electrons; the other is able to produce only muons, and is known as the muon neutrino ν_μ (but see page 170).

Neutrinos, like other particles, have their corresponding

antiparticles, and it is found that the creation and annihilation of electrons and positrons in pairs is a special case of the creation and annihilation of leptons in pairs. This has been represented in a helpful way by the introduction of lepton numbers of two types, one for electrons and their neutrinos (L_e) and the other for muons and muon neutrinos (L_μ), as shown below:

	e^-, ν_e	$e^+, \bar{\nu}_e$	μ^-, ν_μ	μ^+, ν_μ
L_e	+1	-1	0	0
L_μ	0	0	+1	-1

The sum total of L_e, and likewise the sum total of L_μ, is found to be conserved in any interaction, of whatever form, e.g. strong, weak, or electromagnetic. Thus

$$n \rightarrow p + e^- + \bar{\nu}_e + Q$$

$$L_e: \quad 0 \rightarrow 0 + (+1) + (-1);$$

and

$$\gamma \rightarrow e^- + e^+ + Q_\gamma$$

$$L_e: \quad 0 \rightarrow (+1) + (-1);$$

and

$${}^{64}_{29}Cu \rightarrow {}^{64}_{28}Ni + e^+ + \nu_e + Q^+$$

$$L_e: \quad 0 \rightarrow 0 + (-1) + (+1).$$

Similarly for the muon lepton number,

$$\pi^- \rightarrow \mu^- + \bar{\nu}_\mu + Q_\pi$$

$$L_\mu: \quad 0 \rightarrow (+1) + (-1);$$

and for both,

$$\mu^+ \rightarrow e^+ + \nu_e + \bar{\nu}_\mu + Q_\mu$$

$$L_e: \quad 0 \rightarrow (-1) + (+1) + 0$$

$$L_\mu: \quad (-1) \rightarrow 0 + 0 + (-1).$$

The muon has been the subject of some quite exceptionally accurate measurements of the ratio of its magnetic moment to its intrinsic spin. The aim is to test the predictions of quantum electro-dynamics, which have been found to be 'perfect' for electrons within very small limits of error in studies of the energies and intensities of spectral lines for atomic hydrogen.

It is common to express the so-called gyromagnetic ratio as

$$g = \frac{\mu}{I} = \frac{\text{magnetic moment in units of } e\hbar/2m_\mu}{\text{intrinsic spin angular momentum in units of } \hbar},$$

and for both electrons and muons g is very close to 2. (The unit $e\hbar/2m_\mu$ is the magnetic moment of a particle of charge e and mass m_μ moving in a circle with angular momentum $\hbar$, neglecting any effects due to its intrinsic spin and magnetic moment). An experiment at CERN, Geneva, measured $(g - 2)/2$ for the muon and obtained 0.001162 (±5 in the last place). The theoretical value is 0.001165, and this remarkable agreement is viewed as strong support for the validity of quantum electrodynamics. The absence of an 'anomalous' magnetic moment indicates that, like the electron, the muon is subject to weak and electromagnetic interactions (and gravity) but not to the strong interaction.

HADRONS

The term 'hadron' refers to all types of particles that interact with each other through the strong interaction. These particles are subdivided into baryons, which act as 'sources' for the mesons, which are the nuclear field quanta. So far we have encountered the proton and neutron and their antiparticles, and the pi-meson or pion, with a few references to kaons.

Pions were first observed (Lattes, Occhialini, and Powell, 1947) in nuclear emulsions exposed for several months to cosmic rays on a high mountain. They exhibited a decay into muons of a definite energy, indicating a two-body decay, now known to be

$$\pi^- \rightarrow \mu^- + \bar{\nu}_\mu + 33.9 \text{ MeV}$$

and

$$\pi^+ \rightarrow \mu^+ + \nu_\mu + 33.9 \text{ MeV}.$$

In the following year these pions were produced artificially by Lattes and Gardner by means of 380 MeV alpha particles (from the Berkeley synchrocyclotron) impinging on carbon and other targets, and were again detected by nuclear emulsions.

In 1947 Oppenheimer suggested that a neutral pi-meson, postulated by several workers as consistent with Yukawa's theory, would decay with a very short mean life into two gamma photons. In 1950 such energetic photons, of approximately 70 MeV, were observed from targets bombarded by 180-350 MeV protons and also from the interactions of 330 MeV gamma rays derived from the Berkeley electron synchrotron. In 1951 the neutral-pion (π^o) rest mass was accurately deduced from the study of negative pion absorption in hydrogen which is either 'radiative absorption',

$$\pi^- + p \rightarrow n + \gamma + Q \; (E_\gamma \simeq 131 \text{ MeV})$$

or 'mesic absorption'

$$\pi^- + p \rightarrow n + \pi^o + Q$$
$$\pi^o \longrightarrow 2\gamma \; (E_\gamma \simeq 55 - 85 \text{ MeV}).$$

The 30 MeV spread of gamma energies arises from the Doppler shift due to the velocity of the decaying π^o, towards or away from the observer and leads to a value of the mass difference of the pions $m(\pi^-) - m(\pi^o)$ of 4.6 MeV/c^2. Since $m(\pi^-) = 139.6$ MeV/c^2, we have $m(\pi^o) = 135.0$ MeV/c^2. The decay into two gamma photons, and not three, also allows a quantum-mechanical argument for zero spin of the π^0.

The identification of a 'new' particle requires at least one parameter that is different from that of other particles. For many new particles this was simply the mass, although caution is needed in comparing, for example, the masses of charged and uncharged pions or even of the proton and neutron. These same particles also warn against labelling particles as 'new' or 'different' on the basis of different observed mean lifetimes. On the other hand, differences of spin J invariably indicate different particles, and so do large differences of mass.

In the year of the pion (1947) Rochester and Butler observed a pair of tracks forming a V in a cloud chamber exposed to cosmic rays, and deduced that they arose from the decay of a neutral V-particle (now known as the K^0). Since the tracks appeared to be two pions they deduced

$$K^0 \rightarrow \pi^+ + \pi^- + Q \ ,$$

with the mass of the K^0 about $800m_e$ (later found to be about $975m_e$). In 1949, using nuclear emulsions, co-workers of Powell at Bristol identified the K^+ in a decay to three pions,

$$K^+ \rightarrow \pi^+ + \pi^+ + \pi^- + Q \ ,$$

which allowed a fairly accurate determination of the K^+ mass.

In 1951 a Manchester group discovered the first hyperon (i.e. baryon heavier than the neutron) by another V-shaped track, this time identifying the decay particles as a proton and a pion. The neutral parent is now known as a Λ^0 (lambda zero) of mass 1115.4 MeV/c^2, and the decay is

$$\Lambda^0 \rightarrow p + \pi^- + Q.$$

By 1953 many more particles had been discovered in cosmic ray events notably the Σ (sigma) and Ξ (xi) hyperons, and about the same time developments in accelerators allowed more careful determinations of the particle masses, spins, lifetimes, and decay modes. A list is provided in Table 8.1, together with details of other properties, such as parity, isospin, and hypercharge (or strangeness) that are introduced on pages 158 to 161.

Table 8.1.

Type	Symbol and charge	Mass (Me V/c^2)	Mean life (s)	Principal decays	Spin and parity	Isospin	Hypercharge
	γ	0	∞		1^-	0	0
Leptons	ν_e	0	∞		$\frac{1}{2}$		0
	$e\pm$	0·511	∞		$\frac{1}{2}$		0
	ν_μ	0	∞		$\frac{1}{2}$		0
	$\mu\pm$	105·7	$2{\cdot}2 \times 10^{-6}$	$e\nu_e\nu_\mu$	$\frac{1}{2}$		0
Mesons	$\pi^\pm$	139·6	$2{\cdot}6 \times 10^{-8}$	$\mu\nu$	0^-	1	0
	π^0	135·0	9×10^{-17}	$\gamma\gamma$	0^-	1	0
	$K^\pm$	493·8	$1{\cdot}2 \times 10^{-8}$	$\mu\nu, \pi\pi$	0^-	$\frac{1}{2}$	±1
	K^0	497·8		50 per cent K^0_S 50 per cent K^0_L	0^-	$\frac{1}{2}$	±1
	K^0_S	497·8	$8{\cdot}6 \times 10^{-11}$	$\pi\pi$	0^-	$\frac{1}{2}$	±1
	K^0_L	497·8	$5{\cdot}3 \times 10^{-8}$	$\pi\pi\pi, \pi\mu\nu$ $\pi e\nu$	0^-	$\frac{1}{2}$	
	η^0	548·8	3×10^{-19}	$\gamma\gamma, \pi\pi\pi$	0^-	0	0
Baryons	p	938·3	∞		$\frac{1}{2}^+$	$\frac{1}{2}$	1
	n	939·6	960	$pe\nu$	$\frac{1}{2}^+$	$\frac{1}{2}$	1
	Λ^0	1115	$2{\cdot}5 \times 10^{-10}$	$p\pi, n\pi$	$\frac{1}{2}^+$	0	0
	Σ^+	1189	8×10^{-11}	$p\pi, n\pi$	$\frac{1}{2}^+$	1	0
	Σ^0	1193	$< 10^{-14}$	$\Lambda\gamma$	$\frac{1}{2}^+$	1	0
	Σ^-	1197	$1{\cdot}6 \times 10^{-10}$	$n\pi$	$\frac{1}{2}^+$	1	0
	Ξ^0	1315	3×10^{-10}	$\Lambda\pi$	$\frac{1}{2}^+$	$\frac{1}{2}$	-1
	Ξ^-	1321	$1{\cdot}6 \times 10^{-10}$	$\Lambda\pi$	$\frac{1}{2}^+$	$\frac{1}{2}$	-1
	Ω^-	1672	$1{\cdot}3 \times 10^{-10}$	$\Xi\pi, \Lambda K$	$\frac{3}{2}^+$	0	-2

The last of these particles to be discovered was the Ω^- in 1964 (see pages 165 f). Since 1974 a series of so-called 'charmed' particles have been found (see pages 170 ff).

PARTICLE RESONANCES

Most of the hadrons mentioned above have mean lifetimes between 10^{-8} s and 10^{-10} s (exceptions are the two electromagnetic decays of π^0 and Σ^0, and those of the η_0 (3×10^{-19} s),

$$\pi^0 \rightarrow 2\gamma(2 \times 10^{-16}\text{ s}) \text{ and } \Sigma^0 \rightarrow \Lambda^0 + \gamma(10^{-14}\text{ s}),$$

and of course the proton (stable) and neutron (10^3 s)). Such lifetimes are long enough to allow these hadrons to be formed in nuclear collisions, move away from their point of origin, and subsequently decay. The characteristics of the corresponding tracks are fairly obvious, making allowance for 'gaps' where a neutral particle leaves no track. A whole new range of 'particles' has now been discovered with very much shorter lifetimes. They require quite different techniques for study, the first of which, 'resonance scattering', was investigated as early as 1952, although its significance was not realized until higher energies allowed 'resonance production' in 1961. The latter is perhaps easier to follow than the former.

Consider the reaction A(a, b)B. It is clear that the definite energies, kinetic and rest-mass, before the reaction in conjunction with linear-momentum conservation require the two final particles (assumed to be stable) to have definite kinetic energies. On the other hand, a reaction between two initial particles leading to three stable particles A(a, bc)B allows the kinetic energy and linear momentum to be shared in an infinite variety of ways, provided the total energy and linear momentum are conserved. If now we consider the case where two of the three final particles live as 'one particle' (b + B) for a finite mean lifetime τ then the resultant 'two-body' decay to b and B will have an energy uncertainty δE associated with it such that $\delta E . \tau \simeq \hbar$. The result is a tendency towards the definite final energies $T_{(b + B)}$ and T_c of the two-body final state, going over to the energy spread of the three-body final state when τ the mean life of the so-called 'resonant particle' happens to be exceedingly short. Since $\hbar = 6.582 \times 10^{-22}$ MeV s, energy spreads of $T_{(b + B)}$ and hence T_c as large as 65 MeV can be obtained for mean lives of $\tau \simeq 10^{-23}$ s, and many particles with such mean lives have been observed.

A most effective way of appreciating the analysis of these energy spreads is by means of a Dalitz plot. An example is shown in Fig. 8.3(a) for the twin reactions

$$K^- + p \rightarrow (\Lambda^0 + \pi^+) + \pi^-$$

and

$$K^- + p \rightarrow (\Lambda^0 + \pi^-) + \pi^+ \quad (T_{K^-} = 758 \text{ MeV}).$$

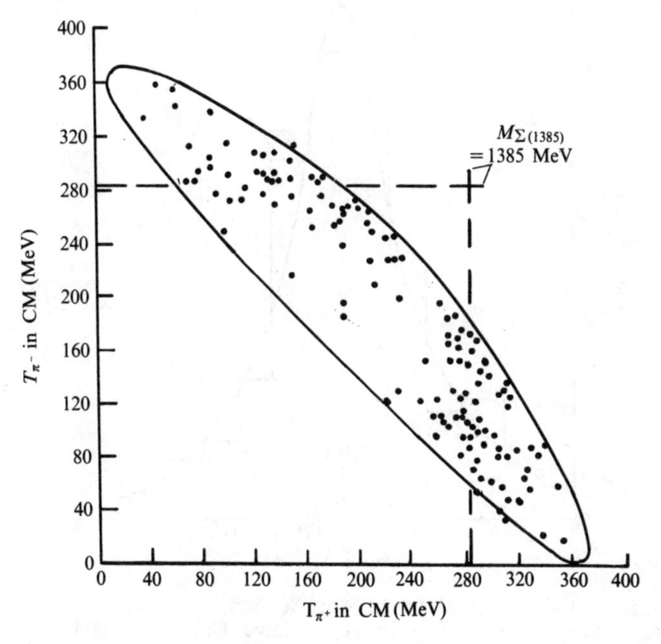

FIG.8.3. (a) Dalitz plot for the Σ (1385) resonances produced by 758 MeV negative kaons on protons. (see Fig. 8.3.(b).)

The resonant particle ($\Lambda^0 + \pi^+$), now known as the $\Sigma^+(1385)$, is formed with a mean mass of 1385 MeV/c^2, and if it were long-lived the kinetic energy of the π^- in the centre-of-mass system would be about 285 MeV. Similarly, for the $\Sigma^-(1385)$ reaction, T_{π^+} would be 285 MeV. For each event observed in a hydrogen bubble chamber the value of T_{π^-} is plotted against T_{π^+} and it is seen that both T_{π^-} and T_{π^+} are spread about the value 285 MeV. If, on the other hand, the reaction had produced three particles (i.e. the lifetime of the $\Sigma(1385)$ had been exceptionally short, say $< 10^{-30}$ s), the coordinates of the plotted points would have been uniformly spread within the envelope of the curve shown, which is set by the joint conservation of energy and linear momentum. For each event a mass of the $\Sigma(1385)$ can be calculated and the result is given in Fig. 8.3(b). Later work has shown

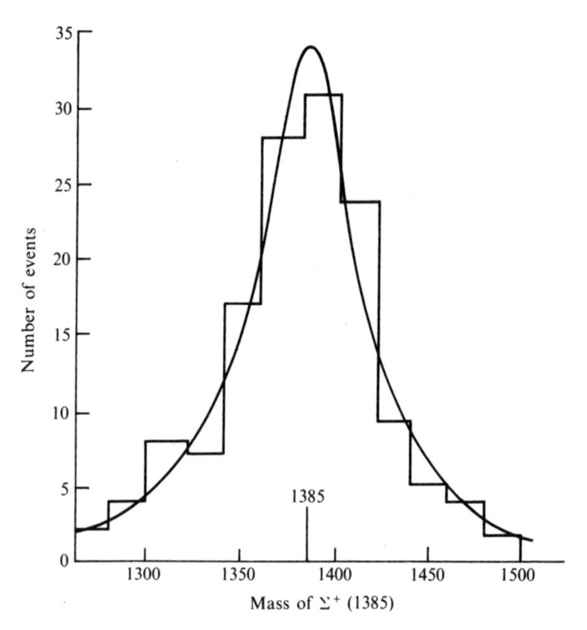

FIG. 8.3. (b) Mass spectrum of the $\Sigma(1385)$ resonances. (Adapted from M.H. Alston et al (1960) Phys. Rev. Lett. 12, 204.)

that the mass is nearer 1382 MeV/c^2 with a mean life of $\sim 1.8 \times 10^{-23}$ s corresponding to $\Delta E \sim 36$ MeV.

Resonance scattering of pions by protons was first recognized in 1952 by Fermi and is indicated by an enhanced probability of scattering near a particular energy of the incident pions (Fig. 8.4). Similar enhanced scattering had been observed for neutrons and interpreted in terms of a compound nucleus of mean life τ given by $\tau\Gamma = \hbar$, where Γ is the width of the resonance, being (in the simplest case) the energy width of the cross-section curve at half its peak value. For the π^+p system there is clearly a resonant state (or particle) of mass 1236 MeV/c^2 with $\Gamma \simeq 120$ MeV/c^2, corresponding to a mean life of 5.5×10^{-24} s. This is known as the $\Delta(1236)$, and from its

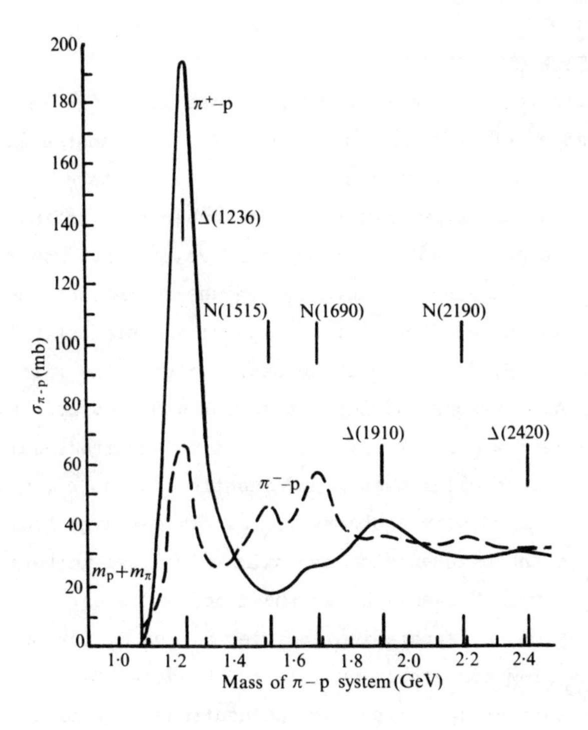

FIG. 8.4. Resonance scattering of π^+ and π^- by protons.

presence also in $\pi^{-}p$ scattering it is clear that it can have positive or negative charge. Fig. 8.4 indicates several more resonant particles. A continuing study of the scattering of different particles (π, K, etc.) over wide ranges of energy have led to the discovery of a great number of resonances. They can be broadly classified into those that decay into baryons and mesons (baryon resonances) and those that decay into mesons only (meson resonances). Notice that some resonances can be observed only in resonance-production reactions and not in resonance scattering. A good example is the $\Sigma(1385)$ discussed above since, in order to be seen in resonance scattering it would require a beam of pions on a Λ^{0} target, or of Λ^{0} on a pion target, and each is manifestly impracticable.

CLASSIFICATION OF HADRONS

Hitherto we have recognized only three identifiable characteristics of hadrons: their mass, electric charge Q, and spin J. For leptons we introduced the lepton numbers L_e and L_μ, and experience has shown that a baryon number B is definable, so that it too is conserved in all types of reactions. Positive values apply to baryons and negative values to antibaryons. $B = 1$ for p and n, and $B = -1$ for $\bar{p}$ and $\bar{n}$; and similarly for the hyperons (Λ, Σ, Ξ, Ω) and antihyperons. No particle with $|B| > 1$ has been found. Mesons and leptons have $B = 0$. Nuclei have $B \geqslant 1$.

The nuclear similarity of the proton and neutron, quite apart from their similar masses and identical spin ($\frac{1}{2}$), has been deduced from many distinct observations. These range from simple facts such as the existence of approximately equal numbers of stable nuclei with Z even - N odd and Z odd - N even, and the near equality of the separation energies S_p and S_n for protons and neutrons from comparable nuclei, to the detailed analysis of high-energy scattering of protons and neutrons. It seems that if we could 'switch off' the electromagnetic interaction the proton and neutron would be indistinguishable. In other words, the neutron and proton appear to be two 'charge states' of the same nuclear particle. As early as 1932 Heisenberg recognized some of the advantages of a formalism parallel to that of the ordinary intrinsic spin of $\frac{1}{2}$ which can be either 'up' or 'down'. Wigner (1937) developed this idea and assigned an 'isospin' of $T = \frac{1}{2}$ for the nucleon, with the values $+\frac{1}{2}$ for the proton and $-\frac{1}{2}$ for the neutron (or more accurately $T_3 = +\frac{1}{2}$ for the proton, and $T_3 = -\frac{1}{2}$ for the neutron, where T_3 is the third component of isospin - in exact analogy to the 'z-component' of ordinary spin). The proton and neutron form an isospin doublet, and other particles fit into other patterns: singlets ($T = 0$), triplets ($T = 1$), and quadruplets ($T = \frac{3}{2}$). Closer inspection reveals that the electric charge $Q = T_3 + \frac{1}{2} B$ for nucleons and for pions, provided that $T = 1$ for the latter (thus T_3 of $\pi^- = -1$, of $\pi^0 = 0$, and of $\pi^+ = +1$, since $B = 0$ for pions). However, for some other particles this formula appears not to be valid. Thus for the Λ^0 which decays to $p + \pi^-$ or to $n + \pi^0$, it is evident that $B = 1$, but since the Λ^0 is equally obviously an isospin singlet it follows that $T = T_3 = 0$, suggesting $Q = \frac{1}{2}$! This was one of a number of problems encountered in studies of hyperons and kaons.

The probability of Λ^0 production by π^- bombardment of protons is quite high (e.g. 10 mb cross-sections are found), and it is clear that the strong interaction is involved, i.e. production is in times of $\sim 10^{-23}$ s. But the decays with mean lifetimes $\sim 10^{-10}$ s are some 10^{13} times as slow, i.e. decay is by the weak interaction, and this is very strange. As Wilkinson remarks, 'So long a lifetime appears to be a suspension of the normal processes of nature and is quantitatively even worse than if Cleopatra had fallen off her barge in 30 B.C. and had taken until last week to splash into the Nile, which would have been too long by a factor of only 10^{11}.' A clue to the solution was observed by Nambu (1951) and Pais (1952), namely, that the 'strange' particles, as they came to be called, were produced in association with each other, although of course they decay in isolation, e.g.

$$\pi^- + p \rightarrow \Lambda^0 + K^0.$$

In 1952 Gell-Mann and, independently, Nishijima accounted for these and other phenomena by the introduction of a new quantum number initially designated 'strangeness' S. (It is part of the fun of physicists, especially in this field, to choose amusing nomenclature. Another quantum number is now well established and is known as 'charm' (see page 170).) For strong and electromagnetic interactions strangeness is conserved ($\Delta S = 0$), but for weak interactions it is not conserved and indeed $|\Delta S| = 1$. Systematic assignments starting with the arbitrary values S = 0 (appropriately enough) for proton and neutron, and opposite values for Λ^0 (S = -1) and K^0 (S = +1), led eventually to a coherent scheme, thus,

production

$$\pi^- + p \rightarrow \Lambda^0 + K^0 \qquad (\Delta S = 0)$$

decay (e.g.)

$$\Lambda^0 \rightarrow p + \pi^- \qquad (\Delta S = +1)$$

and (e.g.)

$$K^0 \rightarrow \pi^+ + \pi^- \qquad (\Delta S = -1).$$

A new expression now links electric charge and isospin, namely, $Q = T_3 + \frac{1}{2} B + \frac{1}{2} S$, leading to the alternative nomenclature $Q = T_3 + \frac{1}{2} Y$, where Y = B + S and is termed 'hypercharge'.

An interesting and typical consequence of the development of the Gell-Mann scheme was the prediction of a 'new' particle. Σ^+ and Σ^- had been observed, but not Σ^0, suggesting $T = \frac{1}{2}$ for the Σ. Decays such as $\Sigma^+ \rightarrow \pi^+ + n$ indicate $B = 1$ and $|S| = 1$, and therefore suggest $Q = +\frac{1}{2}$ or $\frac{3}{2}$, which was not observed. A solution is found if Σ^0 exists, since then $T = 1$ and $S = -1$ satisfies all requirements with $T_3 = +1, 0, -1$ for Σ^+, Σ^0, and Σ^-, respectively. The experimenters duly obliged in 1957 by discovering the Σ^0 in analyses of π^- reactions in a hydrogen bubble chamber. Since $\Sigma^0 \rightarrow \Lambda^0 + \gamma$, $\Delta S = 0$, and the reaction is therefore relatively quick by means of the electromagnetic interaction ($\tau < 10^{-14}$ s). It was rather obviously difficult to observe because of the neutral particles, the full process being

$$\pi^- + p \rightarrow \Sigma^0 + K^0$$
$$\hookrightarrow \Lambda^0 + \gamma$$

Our identifying characteristics for hadrons are now five: mass, spin, baryon number, isospin, and hypercharge. One more is available from quantum mechanics, namely, the intrinsic parity of the wavefunction representing the particle. Parity is said to be positive if the sign of the wavefunction is unchanged as a result of reversing all the coordinate axes, $x \rightarrow -x$, $y \rightarrow -y$, $z \rightarrow -z$. This is equivalent to a mirror reflection followed by a 180° rotation about the perpendicular to the mirror. If the sign of the wavefunction is changed under this transformation, the parity is negative. Only changes of parity can be deduced from observations, coupled with quantum-mechanical considerations, and we choose the intrinsic parity of the proton to be positive. It is common to combine spin J and parity π in a single symbol J^π. Thus for a proton $J^\pi = \frac{1}{2}^+$, and for a pion 0^-. For many years it was assumed that the total parity of a system is conserved for all interactions, but in 1956 Lee and Yang realized that non-conservation of parity in the weak interaction could explain some curious features of the decays of kaons, and they proposed experimental tests using the beta emission of ^{60}Co. These experiments quickly proved the non-conservation of parity under these conditions, but it is still believed to be conserved in the strong and electromagnetic interactions. One conclusion from the experiments is that the neutrino spin and linear-momentum vectors are oppositely directed, as for a left-handed screw. This led Heisenberg to quip that 'God must be a weak left-hander.'

A preliminary classification is now possible. For baryons $B = 1$ we have the 'families' shown in Table 8.2 which include not only the nucleons and hyperons but also the particle resonances.

TABLE 8.2

Baryon Family		Hypercharge (Y)	Isospin (T)
Nucleon	N	1	$\frac{1}{2}$
Lambda	Λ	0	0
Sigma	Σ	0	1
Xi	Ξ	-1	$\frac{1}{2}$
Omega	Ω	-2	0
Delta	Δ	1	$\frac{3}{2}$

For mesons $(B = 0)$ there are three 'families' characterized as π with $Y = 0$, $T = 1$; η with $Y = 0$, $T = 0$; and the rather complicated K with $Y = +1$, $T = \frac{1}{2}$. (We shall not pursue the fascinating story of the kaons because of its numerous facets and the range of new quantum-mechanical concepts involved. It is probable that there is still much to be learned from the kaons, especially the neutral kaons (see Table 8.1), including information on new types of interactions.)

The introduction of isospin enabled particles to be grouped

together in a unifying way, the small differences in masses being accounted for by the 'splitting' produced by the electromagnetic interaction. It is found to be profitable to try to relate certain isospin multiplets of much greater mass differences in larger groupings, with the thought that other interactions, possibly new ones, can account for the 'mass splitting'. The mathematical formalism suitable for isospin multiplets is termed SU(2), meaning 'special unitary group for arrays of size 2 × 2', and an extension to 3 × 3 arrays (i.e. SU(3)) deals with the three dimensions of isospin, the hypercharge dimension, and four new 'spin' dimensions. In 1960 this extension was tried with remarkable success by Gell-Mann, and independently by Ne'eman, first for groups (or so-called 'supermultiplets') of eight baryons, and of eight mesons, and later for a group of ten baryons, and for singlets.

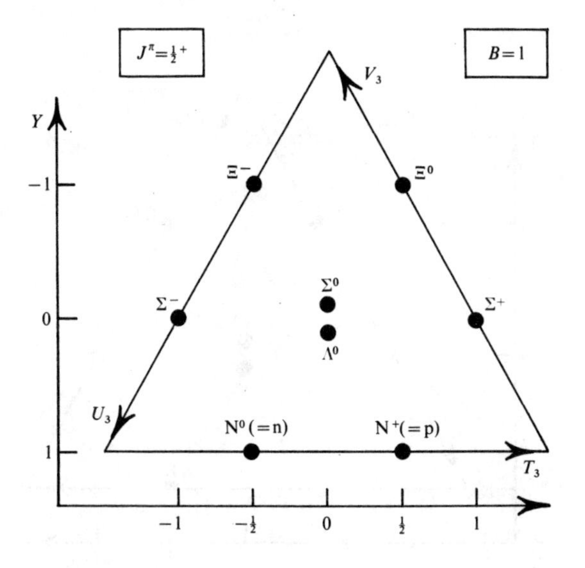

FIG. 8.5. Classification of baryons of $J^\pi = \frac{1}{2}^+$ suggested by unitary symmetry SU(3). This octet includes the proton and neutron (it is drawn in relation to a triangle rather than a hexagon to emphasize the threefold symmetry).

In figure 8.5 the particles with $J^{\pi} = \frac{1}{2}^{+}$ are displayed as functions of their hypercharge Y and isospin T_3; they form an octet. The predicted relation between the masses works well,

$$2\ M_N + 2\ M_{\Xi} = 3\ M_{\Lambda} + M_{\Sigma}\ ,$$

and it is evident that the relation could have been displayed in a closely similar manner by successive rotations through 120°, so that the lines of symmetry labelled U and V successively take the place of the isospin ordinate T_3. The interpretation of so-called U-spin and V-spin remains to be made.

For the familiar group of mesons with $J^{\pi} = 0^{-}$ the pattern is shown in Fig. 8.6, and is also an octet, but in 1960 the η^{0} had not been observed. It was therefore sought by the experimentalists, its mass having been estimated as 567 MeV/c^2

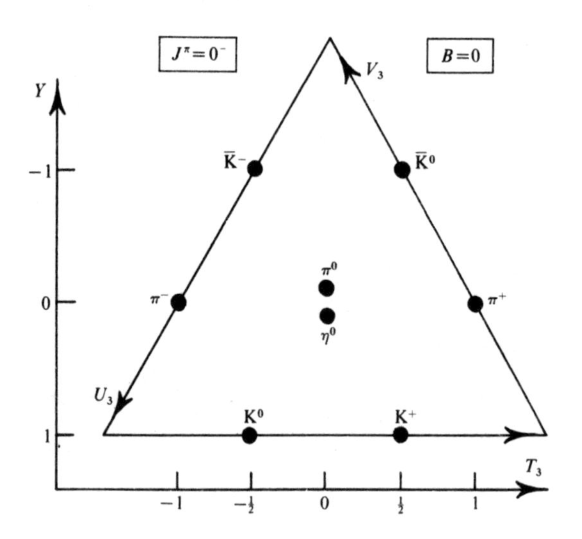

FIG. 8.6. Octet of mesons of $J^{\pi} = 0^{-}$, which led to the prediction of the η^{0}.

from the predicted relation

$$2\,M_K^{\;2} + 2\,M_{\bar{K}}^{\;2} = 3\,M_\eta^{\;2} + M_\pi^{\;2}\,.$$

In 1961, the η^0 was found, with a mass of 549 MeV/c^2 and a mean lifetime of about 3×10^{-19} s - sufficiently close to the predictions to encourage the theoreticians (in 1962 Sakurai interpreted the discrepancy of mass as partial 'mixing' between supermultiplets).

A more striking prediction of SU(3) theory followed from the baryon decuplet (Fig. 8.7) with $J^\pi = \frac{3}{2}^+$. In 1961 the Ω^- had not

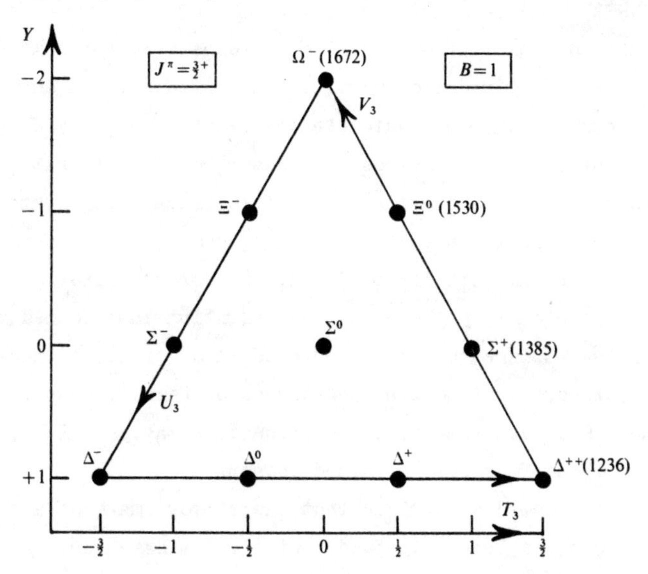

FIG. 8.7. Decuplet of baryons of $J^\pi = \frac{3}{2}^+$ which led to the prediction of the Ω^-.

been found in spite of its 'long' life ($\sim 10^{-10}$ s), although its properties could be predicted with some confidence. Its expected mass of 1675 MeV/c^2 followed from

$$M_\Omega - M_\Xi = M_\Xi - M_\Sigma = M_\Sigma - M_\Delta$$

its electric charge would be -1 from $Q = T_3 + \frac{1}{2} Y$, and its hypercharge of -2 (strangeness -3) implied that its decay would be to $\Xi^0 + \pi^-$, $\Xi^- + \pi^0$, or $\Lambda^0 + K^-$. After a great deal of searching in hydrogen-bubble-chamber pictures of interactions of K^- ($Y = -1$) mesons, the Ω^- was discovered in 1964 (Fig. 5.6) with a mass of 1672 MeV/c^2. The event was hailed as a triumph for the SU(3) theory.

CONSERVATION LAWS AND SYMMETRY

The conservation 'laws' of physics were for some 300 years restricted to those of mass, linear momentum, and angular momentum. In 1905 Einstein's theory of relativity developed the concept of mass-energy as the conserved quantity rather than mass alone. The conservation of electric charge, in integral multiples of the electronic charge, came to be recognized, without being formally proposed, and for the first half of the twentieth century these four conservation laws formed the basis for the whole structure of physical laws or quantifiable relations. Just before the success of Fermi's neutrino theory there had been some doubts expressed, notably by Bohr, but all-in-all the four laws remained intact.

It seemed so obvious that the mirror image of a physical system would behave in fundamentally the same sort of way as the original system that the concept of parity conservation in quantum-mechanical treatments was assumed almost without question. The non-conservation of parity (1955) in the weak interaction came as something of a shock, and led to a more searching investigation of the underlying assumptions in other conservation laws.

It had been known for some time that conservation of mass-energy followed as a deducible necessity from the observation that the behaviour of a given physical system is the same today as it was yesterday, i.e. that it is independent of translation in time. Correspondingly, independence of a translation in space implies conservation of linear momentum (Appendix C), and independence of rotation through a given angle in space implies conservation of angular momentum. Conservation of electric charge cannot be related to such commonsense 'symmetry operations', but is fully understood in terms of so-called gauge invariance (i.e. electric and magnetic fields are independent of shifts in the zero values of the scalar and vector electromagnetic potentials V and A).

How about the conservation of lepton numbers L_e and L_μ, and the baryon number B? So far we have no known symmetry operations that imply these three conservation laws. It is a great challenge to try to identify those 'obvious' unconscious assumptions about the nature of the physical world that imply the conservation of L_e, and L_μ, and B.

Other symmetry operations that have been already identified include 'time-reversal', and 'charge-conjugation'. If a cine-film of a physical event can be run backwards and the time-reversed events follow known physical laws it is clearly impossible, in principle, for an independent observer to know whether the film is running forwards or backwards. This is a highly simplified presentation of what is known as the conservation of 'time-parity' T. Charge-conjugation is the act of symmetry operation in which every particle in a system is replaced by its antiparticle. If the anti-system, or anti-matter counterpart, exhibits the same physical phenomena then 'charge-parity' C is conserved. In fact C (like space-parity P) is not conserved in the weak interaction (a deduction from the right- and left-handedness respectively of positrons and electrons from the decay of μ^+ and μ^-).

The combined symmetry operation in which the anti-matter

mirror image of a system is run in reverse allows a test of CPT invariance, first postulated in 1953 by Schwinger, one predicted consequence of which is that particles and their antiparticles have identical masses and lifetimes. All the evidence, some of high precision, supports the conservation of CPT. On the other hand there is good reason to believe, from observations of decay modes of neutral kaons (1964), that conservation of the combined parities CP does not hold in weak interactions, and if CPT is conserved then T cannot be conserved in weak interactions. The present position is summarized in Table 8.3, where a tick implies conservation, a cross maximum non-conservation, and a question mark indicates partial or probable non-conservation.

TABLE 8.3

	Strong	Electromagnetic	Weak
CPT	✓	✓	✓
P	✓	✓	✗
C	✓	✓	✗
CP or $\dot{T}$	✓	✓	?

ALPHA AND OMEGA AND BEYOND - THE DISCOVERY OF CHARM

The naming by Rutherford of the alpha and beta rays or particles was doubtless based on a justifiable belief in his priority of identification of these two species of radiation. It would be nice to know whether or not the implied finality of the symbol Ω for the last 'long-lived' (1.3×10^{-10} s) hyperon to be discovered was deliberate. No other such hyperon had then been predicted. But now there seems to be the possibility of the identification of even more particle resonances than the hundreds already known, especially as higher energies and more exotic particle beams become more readily available. There are still some 'predictions' of considered value that have not yet been confirmed, ranging from the field quantum of the weak interaction known as the W-particle or intermediate vector boson, through the curious Dirac magnetic monopole, to the much-hunted quarks, partons, and the like.

The quark theory, tentatively formulated by Sakata in 1956, was put in the context of SU(3) group representations by Gell-Mann, and by Ne'eman in 1961. In its first form it suggested that all hadrons are composed of groups of particles selected from six basic quarks, all with spin $\frac{1}{2}$, three particles (q_p, q_n, q_Λ) and three anti-particles (Table 8.4). Mesons would

TABLE 8.4

Quark	B	T	T_3	Y	S	C	Q
q_p, u	$\frac{1}{3}$	$\frac{1}{2}$	$\frac{1}{2}$	$\frac{1}{3}$	0	0	$\frac{2}{3}$
q_n, d	$\frac{1}{3}$	$\frac{1}{2}$	$-\frac{1}{2}$	$\frac{1}{3}$	0	0	$-\frac{1}{3}$
q_Λ, s	$\frac{1}{3}$	0	0	$-\frac{2}{3}$	-1	0	$-\frac{1}{3}$
q_c, c	$\frac{1}{3}$	$\frac{1}{2}$	$\frac{1}{2}$	$\frac{1}{3}$	0	1	$\frac{2}{3}$
$q_{\bar{p}}$, $\bar{u}$	$-\frac{1}{3}$	$\frac{1}{2}$	$-\frac{1}{2}$	$-\frac{1}{3}$	0	0	$-\frac{2}{3}$
$q_{\bar{n}}$, $\bar{d}$	$-\frac{1}{3}$	$\frac{1}{2}$	$\frac{1}{2}$	$-\frac{1}{3}$	0	0	$\frac{1}{3}$
$q_{\bar{\Lambda}}$, $\bar{s}$	$-\frac{1}{3}$	0	0	$\frac{2}{3}$	+1	0	$\frac{1}{3}$
$q_{\bar{c}}$, $\bar{c}$	$-\frac{1}{3}$	$\frac{1}{2}$	$-\frac{1}{2}$	$-\frac{1}{3}$	0	-1	$-\frac{2}{3}$

be constructed from quark-antiquark pairs (e.g. $K^+ = (q_p, q_{\bar{\Lambda}})$) and baryons from three quarks (e.g. $p = q_p, q_p, q_n$)). One striking property of such quarks, if they could be detected, would be their fractional electric charges, $\frac{1}{3}e$ or $\frac{2}{3}e$. Their masses could be large, with compensating large binding energies. In spite of many searches, free quarks have not been observed, but the scattering of very high energy electrons by protons appears to lend support to a proton structure that includes point-like particles (partons) of spin $\frac{1}{2}$.

The search for unity in physical explanations has frequently been rewarded by great and beautiful discoveries. Maxwell's electromagnetic equations are an example, de Broglie's hypothesis another, and SU(3) and quarks perhaps a third. The desirability of unifying two or more of the four well-established fundamental interactions has long been recognized, and discoveries in particle physics are already beginning to provide the necessary evidence. In 1973 experiments at CERN showed that neutrinos interacting with protons need not always convert to charged leptons, as had hitherto been thought. The interpretation in terms of field bosons (page 131, Table 7.1) is that not only charged W-particles but neutral intermediate bosons, the Z^0 particles of Weinberg, communicate the weak force between the neutrino and the proton.

This discovery of so-called neutral currents supported theoretical speculations relating the electromagnetic and the weak interactions, and suggested that many other neutral current interactions should be observed, e.g. $K^0 \rightarrow \mu^+ + \mu^-$, and at appreciable rates. In the same way that the long lives of the hyperons and kaons were explained by introducing a new quantised property called strangeness (or hypercharge, page 160), so now the absence of $K^0 \rightarrow \mu^+ + \mu^-$ and other decays was explained by Glashow by postulating the property of 'charm' (or rather reviving it from less successful proposals in 1964), with a corresponding charm quantum number and a new conservation law.

The next step was to introduce charmed quarks, q_c, with $T = \frac{1}{2}$, $Y = \frac{1}{3}$, $B = \frac{1}{3}$, $S = 0$, $Q = \frac{2}{3}$ and corresponding antiquarks,

$q_{\bar{c}}$. (Alternative quark symbols (Table 8.4) are $q_p = u$ (for 'up'), $q_n = d$ (down), $q_\Lambda = s$ (strange) and $q_c = c$ (charmed).) The charm number is 1 for charmed quarks, -1 for charmed antiquarks and 0 for u, d and s quarks and antiquarks. It thereby became possible to construct hypothetical particles with distinctive new properties, e.g. $(c, \bar{c})$ mesons with zero total charm which in decay may reveal 'hidden charm', and $(c, \bar{s})$ with non-zero total charm, sometimes referred to as 'naked charm'.

Once again theoretical speculations were rewarded by significant experimental discoveries. In late 1974, a neutral particle of mass 3100 MeV/c^2 was discovered at Brookhaven during 29 GeV proton bombardment of beryllium. Independently, and almost simultaneously, it was identified in electron-positron collisions in intersecting storage rings at Stanford where, within a few weeks, a 3690 and, later, 4100 and 4400 MeV/c^2 particles were also found. The lifetimes (10^{-20} s) of the two lighter of these so-called 'psi-particles' (ψ) (also termed J-particles) are about 1000 times longer than those for typical resonances, corresponding to uncertainties in mass of less than 1 MeV/c^2. In 1975, a team at Hamburg deduced the existence of a short-lived 3520 MeV/c^2 particle formed by photon emission from $\psi(3690)$ and decaying by a γ to $\psi(3100)$. This was later confirmed at Stanford, and labelled $\chi(3520)$, together with $\chi(3410)$ formed by γ-emission from $\psi(3690)$ and decaying by a γ to $\psi(3100)$. Hamburg also reported $\psi(3100) \rightarrow \gamma + \eta_c(2800)$ and a $\chi(3560)$ has also been mooted.

The introduction of charm requires the two-dimensional relations of Figs. 8.5 - 8.7 to be extended to three dimensions (Fig. 8.8). The $(c,\bar{c})$ particles such as the ψ/J, then appear at the centre, and the charmed particles and anti-particles are out of the Y,T_3 plane. Their masses have been predicted and agree well with experimental values, including the 1860, 2020 and 2120 MeV/c^2 charmed mesons reported in 1976. The close parallel to

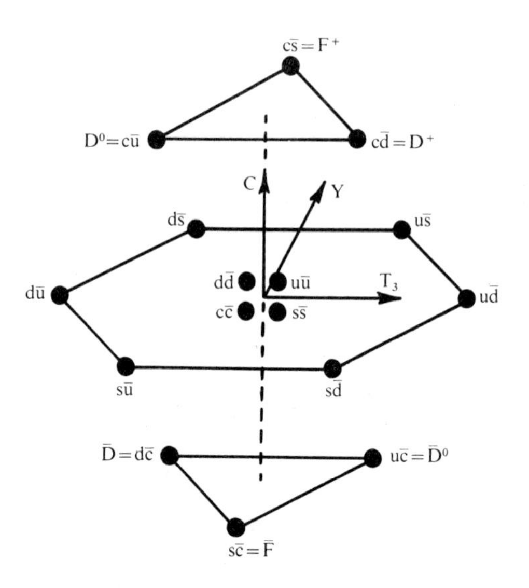

FIG. 8.8. Three dimensional structure of mesons constructed from four quarks and four antiquarks, producing three new charmed mesons D^0, D^+, F^+ and their antiparticles, and the $(c,\bar{c})$ particle believed to be the ψ- (or J-) particle for the case where c and $\bar{c}$ have parallel spins. Excited states of ψ have been identified. It is possible that the 1860 MeV/c^2 particle reported in 1976 is either D or F.

the excited states of positronium has led to the coining of the name 'charmonium' for the ψ-particles. Observed decay modes also agree with predictions and lend support to earlier claims of charmed particles found in bubble-chamber experiments (1976) at CERN (production of $\mu^- e^+ K^0$ by ν_μ beams) and in counter experiments (1976) at Batavia Fermilab (muon pairs produced by ν_μ, believed to be

$$\nu_\mu + \text{nucleus} \rightarrow \mu + C \text{ (charmed)} + \text{nucleus}$$

followed by $C \rightarrow \mu + \nu_\mu$ + other particles). Spark-chambers and nuclear emulsions have jointly produced visual evidence of charmed particles (Europe-USA, Nov. 1976). A charmed baryon of mass 2260 MeV/c^2 and the surprisingly stable upsilon particle of mass

6000 MeV/c^2 have been reported at Fermilab (1976).

Theories involving three different 'colours' for each of the four quarks, and coloured gluons to hold the quarks together in particles of zero net colour, suggest that quarks may be in principle unobservable. Any attempt to isolate a quark would simply result in the creation of a new quark-antiquark meson.

Comparably remarkable is new evidence from Stanford (1977) of electron-muon pairs interpreted as $e^+e^- \rightarrow U^+U^-$ where U is a new 'heavy lepton' of mass about 1900 MeV/c^2 (U = 'unknown') and $U \rightarrow \mu\ \bar{\nu}_\mu\ \nu_U$ or $e\ \bar{\nu}_e\ \nu_U$, with ν_U the corresponding U-neutrino. If confirmed there would then be three lepton numbers to conserve; L_e, L_μ, L_U. Speculations about a symmetry between leptons (e, ν_e), (μ, ν_μ) and quarks (u, d), (s, c) might then suggest (U, ν_U) paralleled by new quarks (t, b). Having discovered 'strange' hadrons and 'charmed' hadrons the experimentalists might then find yet heavier t- and b- hadron families.

There is still room for surprises in the continuing search for unity between the identified interactions of physics, but it now appears almost certain that the electromagnetic and weak interactions have been shown to have an underlying identity. Atomic physics experiments to check this (Oxford 1976) have cast some doubt. Strong interaction unification remains a challenge. Gravity is less easy to investigate and it may be that astronomy and cosmology will provide the evidence and understanding to link the very big and the very small of both space and time.

PROBLEMS

8.1. In Anderson's first cloud-chamber picture of the positron (Fig. 8.2(a)) the radius of the track of the incident particle is 140 mm, and after passing through a 6.0 mm lead plate the track radius is 51 mm. The applied magnetic flux density B was 1.50 T, and the plane of each track was approximately perpendicular to B. (a) Calculate the initial and final energies of the positron and compare its mean rate of energy loss with that for an electron, as shown in Fig. 5.1(b). What can be deduced from the droplet densities of the two tracks? (b) Show that the incident particle could not have been a proton. The semi-empirical range-energy relation $R(m) = \{T(\mathrm{MeV})/9.3\}^{1.8}$ for protons in air, is valid from a few MeV to 200 MeV. Is this relevant to the problem?

8.2. A muon of rest mass 105.7 MeV/c^2, moving in a cloud chamber at right-angles to a magnetic field of flux density 1.00 T, passes through a metal plate. The radii of curvature of its coplanar tracks before and after the plate are 0.50 m and 0.25 m respectively. Find the energy (in MeV) lost in the plate. Compare the answers obtained by the use of relativistic and non-relativistic formulae.

8.3. Show that a photon cannot spontaneously produce an electron-positron pair in the absence of another particle.

8.4. Calculate the energies of the muon and muon-neutrino produced in the decay of a stationary pion.
($M_{\pi+} = 139.6\ \mathrm{MeV}/c^2$, $M_{\mu+} = 105.7\ \mathrm{MeV}/c^2$.)

8.5. Which of the following reactions and decays are 'forbidden' by one or more conservation laws? State the laws violated.

(a) $n \rightarrow p + 2e^- + e^+ + \nu_e$.
(b) $\mu^+ \rightarrow e^+ + \gamma + \bar{\nu}_e$.
(c) $\pi^o \rightarrow \mu^- + e^+$.
(d) $\pi^- \rightarrow \mu^+ + 2e^- + \nu_\mu$.
(e) $p + p \rightarrow \Sigma^+ + K^+$.
(f) $n + d \rightarrow \Sigma^o + \Sigma^o + p$.
(g) $K^- \rightarrow \pi^- + \mu^+ + \mu^-$.
(h) $K^+ \rightarrow e^+ + \pi^o + \nu_e$
(i) $\pi^+ + n \rightarrow \Lambda^o + K^+$
(j) $K^- + p \rightarrow \Xi^o + K^o$

Appendix A: Rutherford Scattering theory

Consider a nucleus of mass M, electric charge Ze, acting as a stationary scattering centre for a parallel beam of particles each of mass m, charge ze, and momentum p. For simplicity we shall at first assume M >> m so that nuclear recoil can be neglected and all the angles can be taken as angles in the laboratory system.

Let the impact parameter d (Fig. A.1) lead to a scattering angle θ, and consider the incident particle at point A. The diagram is clearly symmetric about the line NS since the motion of the particle can be reversed without affecting the result. Let φ be the angle between NS and NA. The force F on the particle at point A is equal in magnitude to $(1/4\pi\varepsilon_0)zZe^2/r^2$, where r = NA. The change of linear momentum during the complete

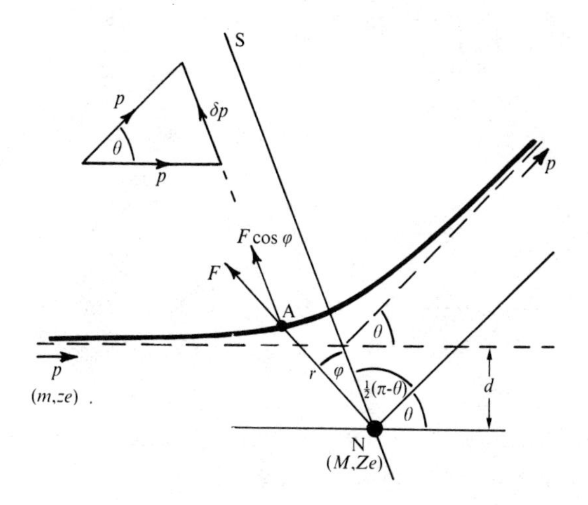

FIG. A.1. Parameters in the Coulomb scattering of a charged particle A by a nucleus N. The impact parameter is the distance d.

scattering process amounts to $\delta p = 2\,p\sin(\theta/2)$ parallel to NS, as can be readily deduced from the vector diagram (Fig. A.1), and is produced by the component of F parallel to NS, i.e. $F\cos\varphi$. Hence

$$\int_0^\infty \left(\frac{1}{4\pi\varepsilon_0}\right)\frac{zZe^2}{r^2}\cos\varphi\,dt = 2\,p\sin(\theta/2).$$

Since angular momentum is conserved,

$$p\,d = m\left(r\frac{d\varphi}{dt}\right)r\ ,$$

and this allows the elimination of r, yielding

$$\frac{zZe^2}{4\pi\varepsilon_0}\int_0^\infty \frac{m}{p\,d}\cos\varphi\,\frac{d\varphi}{dt}\,dt = 2\,p\sin(\theta/2)\ .$$

The limits of φ are $-\tfrac{1}{2}(\pi-\theta)$ and $+\tfrac{1}{2}(\pi-\theta)$, hence

$$\frac{zZe^2}{4\pi\varepsilon_0}\cdot\frac{m}{2\,p^2 d}\int_{-\frac{1}{2}(\pi-\theta)}^{\frac{1}{2}(\pi-\theta)}\cos\varphi\,d\varphi = \sin(\theta/2).$$

It is convenient to introduce b_0, the distance of closest approach for $\theta = 180^\circ$ (Fig. 3.3), given by

$$b_0 = \frac{zZe^2}{4\pi\varepsilon_0}\cdot\frac{2m}{p^2}$$

Then, by integration,

$$\cot(\theta/2) = 2d/b_o.$$

As explained on page 41, the differential scattering cross-section is the probability of scattering by one nucleus per unit area into unit solid angle in the direction θ. Consider therefore a range of the impact parameter $+\delta d$, which will produce a range of $-\delta\theta$ (Fig. A.2); notice that an increase of d leads to

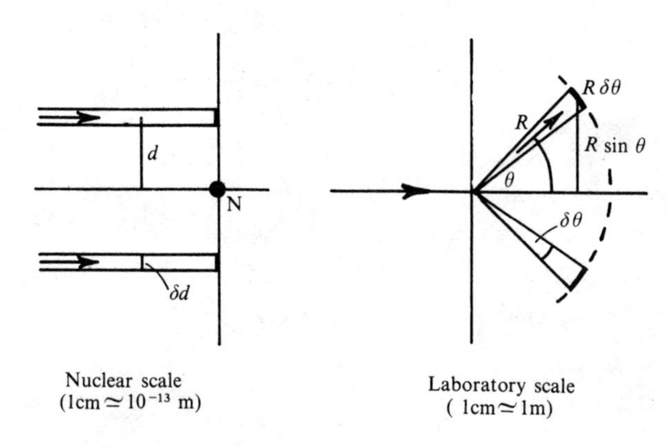

FIG. A.2. Relation between range of impact parameter and range of scattering angle.

a decrease of θ. For one nucleus per unit area, the probability of scatter into $-\delta\theta$ is clearly the ratio of the area $2\pi d.\delta d$ to unity. The solid angle corresponding to $-\delta\theta$ is $2\pi R\sin\theta(-R\delta\theta)/R^2$, i.e. $-2\pi\sin\theta\delta\theta$. Hence

$$\frac{d\sigma(\theta)}{d\Omega} = \frac{-2\pi d \cdot \delta d}{2\pi\sin\theta\,\delta\theta} = \frac{-d}{\sin\theta} \cdot \frac{\delta d}{\delta\theta}.$$

From the differentiation with respect to θ of $\cot(\theta/2) = 2d/b_o$,

$$-\frac{1}{2\sin^2(\theta/2)} = \frac{2}{b_o}\frac{\delta d}{\delta\theta},$$

and hence, using $\sin\theta = 2\sin(\theta/2)\cos(\theta/2)$,

$$\frac{d\sigma(\theta)}{d\Omega} = \frac{-(b_o/2)\cot(\theta/2)}{2\sin(\theta/2)\cos(\theta/2)}\left\{\frac{-b_o/2}{2\sin^2(\theta/2)}\right\}$$

or

$$\frac{d\sigma(\theta)}{d\Omega} = \frac{b_o^{\ 2}}{16\sin^4(\theta/2)}.$$

Allowance for the finite mass of the nucleus involves two corrections, which are small for low-energy (~ 5 MeV) alpha scattering from heavy nuclei. The first is the use of the reduced mass in the expression for b_o,

$$b_o = \frac{zZe^2}{4\pi\varepsilon_o}\cdot\frac{2m}{p^2}\left(\frac{M+m}{M}\right).$$

The second arises from the translation of angles (θ_{CM}) in the centre-of-mass system to angles (θ) in the laboratory system,

$$\left\{\frac{d\sigma(\theta)}{d\Omega}\right\}_{CM} = \frac{b_o^{\ 2}}{16\sin^4(\theta_{CM}/2)},$$

with

$$\tan\theta = \frac{\sin\theta_{CM}}{m/M + \cos\theta_{CM}}$$

Notice that in calculating the number scattered,

$$N(\theta_{CM}) = \left\{\frac{d\sigma(\theta)}{d\Omega}\right\}_{CM} N_o \; nx \; \delta\Omega_{CM} \; ,$$

and since $\delta\Omega = 2\pi \sin\theta \; \delta\theta$,

$$\delta\Omega_{CM} = \left[\frac{\left\{1 + (2\, m/M) \cos\theta_{CM} + m^2/M^2\right\}^{\frac{3}{2}}}{1 + (m/M) \cos\theta_{CM}}\right] \delta\Omega ,$$

with $\delta\Omega = D/R^2$ for a detector of area D at a distance R from the scattering centre.

Appendix B: Transmission through a potential barrier

Consider a beam of particles, each of mass M and kinetic energy T incident on a square potential barrier (Fig. 3.10.(a)) with $V(x) = 0$ for $-b > x > 0$, and $V(x) = V$ for $-b < x < 0$.

In the region $x < -b$ the incident wave can be assigned unit amplitude and is represented by $\exp(i\alpha x)$, and the reflected wave is correspondingly $A \exp(-i\alpha x)$. The total wave in this region is therefore

$$\psi(x) = \exp(i\alpha x) + A \exp(-i\alpha x),$$

and this satisfies the wave equation for $\alpha^2 = 2MT/\hbar^2$, with the amplitude A as yet unspecified and in general complex.

Within the barrier the form of the wavefunction cannot be that of a free particle, and the 'waves' in the two directions are $B \exp(\beta x)$ forwards and $C \exp(-\beta x)$ in reverse, i.e. reflected at $x = 0$. Then $\beta^2 = 2M(V - T)/\hbar^2$, and B and C are in general complex. Hence for $-b < x < 0$

$$\psi_2 = B \exp(\beta x) + C \exp(-\beta x).$$

The transmitted wave is simply

$$\psi_3 = D \exp(i\alpha x),$$

where D is also in general complex and the probability of transmission is $|D|^2$.

The boundary conditions require continuity of both ψ and $d\psi/dx$ at $x = -b$ and $x = 0$, and it follows that we have four equations for the four complex unknowns.

At $x = 0$,

$$B + C = D,$$

$$-\beta B + \beta C = i\alpha D.$$

At $x = -b$,

$$\exp(-i\alpha b) + A\exp(i\alpha b) = B\exp(\beta b) + C\exp(-\beta b)$$

$$i\alpha\exp(-i\alpha b) - Ai\alpha\exp(i\alpha b) = -B\beta\exp(\beta b) + C\beta\exp(-\beta b).$$

For the present purposes only D is needed. Successive elimination of A, B, and C yields, after a little tedious algebra,

$$D = 2i\alpha\exp(-i\alpha b)\left[\frac{\alpha^2 - \beta^2}{2\beta}\{\exp(\beta b) - \exp(-\beta b)\} + i\alpha\{\exp(\beta b) + \exp(-\beta b)\}\right]^{-1}.$$

The transmitted intensity is then

$$|D|^2 = \left(\frac{4\alpha\beta}{\alpha^2 + \beta^2}\right)^2\{\exp(2\beta b) + \exp(-2\beta b) + 2\}^{-1}$$

or

$$|D|^2 = \left(\frac{2\alpha\beta}{\alpha^2 + \beta^2}\right)^2\{1 + \sinh^2(\beta b)\}^{-1}.$$

Inserting the values of α^2 and β^2 leads to

$$|D|^2 = 4(T/V)(1 - T/V)\ \{1 + \sinh^2(\beta b)\}^{-1},$$

with βb for alpha particles given by

$$\beta b = 0.4365b\ (V - T)^{\frac{1}{2}}$$

for b in fm and V and T in MeV.

The forms of the wavefunctions in the three regions $x < -b$, $-b < x < 0$, and $x > 0$, are shown in Fig. B.1, which is drawn to emphasize two points,

(1) that it is not possible to represent the amplitude of the wavefunction in all three regions as a purely real quantity;

(2) that the forms of the wavefunction within the barrier depend on the phase of the incident wave which has a time-dependence $\exp\ (-iTt/\hbar)$.

In a wave-mechanical representation of a stationary state the real part of the wavefunction is always of the same form as the imaginary part but of different amplitude, and it is possible to select a point in time when the imaginary part has zero amplitude, i.e. the wavefunction is purely real throughout the system. Whenever there is a flux of particles, as through a potential barrier, it is not possible to find a point in time at which the wavefunction is purely real (or imaginary) throughout the system, and therefore it is necessary, for a proper interpretation of the wavefunction, to consider both the real and imaginary parts.

The usual diagrams of wavefunctions in potential barriers appear similar to that of the real part of Fig. B.1.(a), probably in order to illustrate that the intensity (i.e. $|\psi|^2$) decreases monotonically within the barrier (Fig. 3.10.(b)). But it is clear that $|\psi|^2$ is the same for all the incident phases shown in Fig. B.1, and the amplitude of the real part (or the imaginary part) can have any of the forms shown, for all of which $d^2\psi/dx^2$ is positive within the barrier.

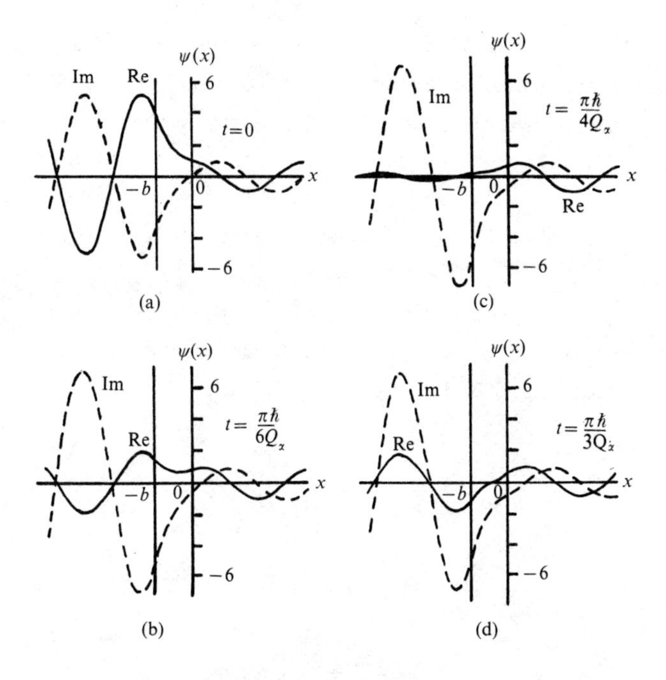

FIG. B.1. Wavefunctions calculated for a beam of particles incident on a potential barrier, at four different times, showing the effect of incident phase on the form of the real (Re) and imaginary (Im) parts of the wavefunction inside the barrier. (The width of the barrier is set at twice the reduced wavelength ($\lambda/2\pi$) of the incident beam; the height of the barrier twice the kinetic energy of the beam; the transmitted intensity is unity; in the expressions for t the kinetic energy is taken as Q_α.)

Appendix C: Linear momentum conservation and linear-translation invariance

We commonly accept, and regularly confirm to great accuracy, that the location in space of an experiment does not influence its result. In other words there is a 'symmetry' based on the two locations or, alternatively, it is impossible to determine an absolute position in space. An important consequence of this 'invariance' is that linear momentum is conserved, as may readily be shown.

Consider a system of two particles, of masses m_1 and m_2, at the points x_1 and x_2. Let F_1 be the force acting on particle m_1 due to particle m_2. By Newton's second law,

$$\frac{d}{dt}(m_1 \dot{x}_1) = F_1,$$

or, in terms of the interaction potential energy function,

$$F_1 = -\frac{\partial}{\partial x_1} V(x_1, x_2)$$

hence

$$\frac{d}{dt}(m_1 \dot{x}_1) = -\frac{\partial}{\partial x_1} V(x_1, x_2),$$

which is the equation of motion for particle m_1.

Now impose a linear translation of the system by a distance ℓ in the x-direction, such that

$$x_1 \to X_1 = x_1 + \ell, \quad x_2 \to X_2 = x_2 + \ell.$$

The new equation of motion is

$$\frac{d}{dt}(m_1\dot{X}_1) = -\frac{\partial}{\partial X_1} V(X_1,X_2),$$

and since

$$\frac{\partial V}{\partial x_1} = \frac{\partial V}{\partial X_1} \cdot \frac{\partial X_1}{\partial x_1}$$

this reduces to

$$\frac{d}{dt}(m_1\dot{x}_1) = -\frac{\partial}{\partial x_1} V(x_1 + \ell,\ x_2 + \ell).$$

For an infinitesimal translation,

$$V(x_1 + \ell,\ x_2 + \ell) = V(x_1,x_2) + \ell\left(\frac{\partial V}{\partial x_1} + \frac{\partial V}{\partial x_2}\right),$$

and hence $(\partial V/\partial x_1 + \partial V/\partial x_2)$ must be zero if the two equations of motion, initial and new, are to be identical. (All potential functions of the form $V(x_1,x_2) = V(x_1 - x_2)$ satisfy this condition, as might be expected since $V(x_1,x_2)$ depends only on the relative coordinates.)

It now follows immediately that

$$\frac{d}{dt}(m_1\dot{x}_1) + \frac{d}{dt}(m_2\dot{x}_2) = -\frac{\partial V}{\partial x_1} \cdot \frac{\partial V}{\partial x_2} = 0\ ,$$

i.e.

$$m_1\dot{x}_1 + m_2\dot{x}_2 = \text{constant},$$

or the linear momentum in the direction x is conserved.

A similar derivation can be made for conservation of angular momentum based on invariance under rotation through a given angle, i.e. it is impossible to measure an absolute direction in space. Likewise conservation of energy arises from invariance under time-translation, i.e. it is impossible to measure an absolute 'point' in time.

Notice the form of this proof. It starts from Newton's second law in order to define linear momentum and uses the 'observed' linear-translation invariance. The conservation of linear momentum, so defined, can also be proved from Newton's third law. The more general principal of linear-translation invariance is preferable for this purpose, although the derivation above, because it is in terms of conservative forces, is also limited in its generality. The conservation law itself is not so limited.

Answers to numerical problems

1.1. $V \leq 2.5$ kV.

1.3. 10^4 s^{-1}.

1.4. $\lambda = 124$ nm, $T_{max} = 5.28$ eV, $v_{max} = 1.36 \times 10^6$ m s^{-1}.

2.1. 1.2×10^{10} times as great, if pure ^{210}Po. 2.8×10^6 (i.e. less) for ^{226}Ra.

2.2. Within 1.5×10^{-7} of complete recovery.

2.5. (a) 1.53×10^{-7} m^3 (0.153 ml); (b) 9.26×10^{-4} ml.

2.6. 6×10^9 years.

2.8. (a) $R_p = 4.3$ cm, $R_d = R_\alpha = 6.12$ cm; (b) $R_p = R_d = R_\alpha = 36$ cm.

2.10. $T/m_oc^2 = \beta^2/2 = 0.06593$ (classical), $T/m_oc^2 = (1 - \beta^2)^{-\frac{1}{2}} - 1$ (relativistic) $= 0.0736$, hence $T_e = 37.4$ keV.

3.1. $B = (2T_\alpha M_\alpha)^{\frac{1}{2}}/ed(2 + 2^{\frac{1}{2}})$, for angle of incidence 45°, $= 3.2 \times 10^5$ T.

3.2. $d\sigma/d\Omega = 6.27$ b sr^{-1}; $N_\theta = 0.028$ s^{-1}.

3.4. $f = A(Zn)Z^2(Cd)/A(Cd)Z^2(Zn)$, $Z(Cd) = 48$.

3.5. (a) 117 keV; (b) 1.7 eV.

4.1. 300 MeV.

4.2. 6.2 MV m^{-1}.

4.3. 4.9 cm.

4.4. 0.55.

4.5. (a) 1.46 m; (b) 0.54 T.

5.1. (a) $T/2$; (b) R proportional to T/M.

5.2. $\Delta T = T_o\{1 - (A^2 + 2A\cos\varphi + 1)/(A + 1)^2\}$,

$$\cos\theta = \frac{A\cos\varphi + 1}{(A^2 + 2A\cos\varphi + 1)^{\frac{1}{2}}} .$$

5.3. $T_e = h\nu\{2\alpha/(1 + 2\alpha)\}$ for $\alpha = h\nu/m_ec^2$, 0.963, 1.119 MeV.

5.4. $d(A\ell)/d(Pb) = 13.13$, 162.4, 4.83, 13.54, 16.74.

5.6. $\Delta V = 5.34$ mV.

5.7. $m_oc^2 = 108$ MeV (i.e. a muon).

5.8. 6.3 ℓ.

5.9. 6202 years

5.10. 6.13 Ci, 36.5°C.

5.11. 111 min.

6.1. $h\nu$ = 2.22 MeV, T_d = 1.32 keV.

6.2. $Q = T_1(m_1/m_4 - 1) + T_3(m_3/m_4 + 1) - 2\{(m_1/m_4)T_1T_3\}^{\frac{1}{2}}$.

6.4. (a) ≃ 10 MeV; (b) 66.927150 u.

6.5. 1.14 MeV.

6.7. e - capture, Q = 0.81 MeV.

6.8. 2.0×10^4 b.

6.10. ≃ 0.14 g.

6.11. 256 MeV for $R = r_o A^{\frac{1}{3}}$ with r_o = 1.2 fm.

6.12. (a) 4×10^9 kg s^{-1}; (b) (i) 9.7×10^{-6}, (ii) 4.9×10^{-2}.

6.13. (a) 103 MeV; (b) 192 MeV, Maximum recoil energy = $4T_o m_1 m_2/(m_1 + m_2)^2$. (Led to discovery of neutron - such observed recoils could not be explained by gamma photons.)

7.3. Kb = 1.831, V_o = 33.8 MeV, $V_o b^2$ = 1.49 MeV b.

7.4. d = 1.22 fm.

7.5. Ratio is $0.70\ Z^2/0.72Z(Z - 1)$.

7.6. B = 1637 MeV, B/A = 7.87 MeV, $B/m_p c^2$ = 1.75, (surface energy)/B = -0.37.

7.7. Q_α = 6.014, 4.843, 3.956, 3.221 MeV.

7.8. $Z = (4a_4 + m_n - m_p)/(8a_4/A + 2a_3 A^{-\frac{1}{3}})$, Z = 8, 19, 35, in agreement with observations.

8.1. (a) 63 MeV, 23 MeV, mean dE/dx ≃ 67 MeV cm^{-1};
(b) T_p = 2.1 MeV, range in air = 7 cm - but no visible change of curvature, and a 2 MeV proton would not pass through 6 mm of lead.

8.2. Relativistic: 183.5 - 129.6 = 53.9 MeV;
non-relativistic: 106.4 - 26.6 = 79.8 MeV.

8.4. T_μ = 4.1 MeV, T_ν = 29.8 MeV, $T_\mu = (m_\pi c^2 - m_\mu c^2)^2/2m_\pi c^2$.

Index

activity, 17ff
alpha particle, 16, 21, 36ff
  absorption, 60
  emission, 53ff
  reactions, 60ff
  scattering, 38ff, 47
  spectra, 50ff
annihilation of
  particle-antiparticle, 145f
antiparticle, 145, 147
atomic number, 8, 20f
  structure, 3ff, 39
Avogadro constant, 4

barn, 42
barrier penetration, 54ff, 114, 180
baryon, 131f
  number, 158
beam characteristics for
  charged particles, 64
Becquerel, Henri, 15ff
becquerel, 20
beta particle, 16, 24, 30, 39, 52f, 102, 130
betatron, 69
binding energy, electron, 33, 101, 134
  nucleon, 33, 107, 112, 124f, 134
black-body radiation, 11f
breeder reactor, 115f
bremsstrahlung, 69, 80, 82
bubble chamber, 90ff
build-up factor, 83

centre-of-mass system, 75, 103, 123, 178
Čerenkov detector, 89
chain reaction, 109, 114
charm, 160, 170
charmonium, 172
charge conjugation, 167
classification of hadrons, 158
cloud chamber, 61, 91, 146, 152
Cockcroft-Walton voltage
  multiplier, 66f, 71
collective model, 137
coloured quarks, 173
compound nucleus, 59, 61, 107, 140, 157
Compton effect, 83
conservation laws, 26, 29, 37, 62, 100f, 158, 160f, 166ff, 168, 184

conversion electron, 25
cosmic rays, 81, 129, 143
Coulomb scattering, 38ff, 43ff,
104, 175
coupling constant, 128, 130
critical size, 115
cross-section,
differential, 40ff, 104, 177
total, 104
Curie, Marie and Pierre, 16ff
curie, 20
cyclotron, 71

delta rays, 81
deuteron, 123ff
diffraction of particles, 47,
49f
direct interactions, 140f
dose,
dose equivalent, 84
maximum permissible, 85
dosimeter, 86

Einstein, Albert, 1, 11, 13,
31, 166
electromagnetic waves, 10, 15,
16
electron, 4, 7f
neutrino, 149
e/m, 6, 24, 30
electron capture, 103
energetics of reactions, 100
energy levels of nucleons, 133
energy loss by nuclear
particles, 79ff
exchange force, 128f
excited states of nuclei, 51,
104ff, 107, 137

Fermi, Enrico, 26, 44, 49,
105, 107
fermi, or femtometer, 44
field quantum, 128, 131
fission, 108, 140
focusing of particle beams,
72ff
fusion, 116

gamma ray, 16, 26, 50, 51,
53, 83, 87
Geiger, Hans, 21, 24, 38ff,
46, 53
Geiger(-Mueller) counter, 22,
86
Geiger-Nuttall relation, 24,
57
Gell-Mann,Murray, 160ff, 169
gluon, 173
graviton, 131f
gray, 83
gyromagnetic ratio, 150

hadron, 131f, 150
half-life, 19
Heisenberg uncertainty relation,
105, 107, 128, 154

hydrogen atom wave functions, 121f
hydrogen bomb, 118
hypercharge, 160
hyperon, 131f, 152

impact parameter, 126, 175
intermediate vector boson, 130, 170
intersecting storage rings, 73, 75
invariance, 166ff, 184ff
ionization by charged particles, 23
ionization detectors,
  gas, 85, 143
  solid, 87
ionization potential, 33, 134
isomer, 93
isospin, 159
isotone, 93
isotope, 29, 93ff

kaon, 130, 152

lepton, 132
  muon lepton number, 149
  electron lepton number, 149
lifetimes of excited states, 105
linear accelerator,
  electron, 69
  proton, 70
liquid drop model, 138, 140

magic numbers, 53, 112, 134f
mass-energy relation, 31, 101
mass spectrograph, 27ff
meson, 129, 131f, 148
moderator, 82, 114
muon, 129, 148
  -neutrino, 149

neutral currents, 170
neutrino, 26, 79, 102, 130, 149, 162, 170
neutron, 62, 79, 82
  delayed, 60, 109
  reactions, 105ff
  separation energy, 134
  thermal, 113
neutron-proton scattering, 42, 122, 125
nuclear bomb, 109, 114
nuclear emulsion, 92, 150ff
nucleon, 112, 158ff

optical model, 48, 104
omega minus, 165, 168

pair production, 83, 145
pairing energy, 113, 140
parity, 104, 161
  charge parity, 167
  time parity, 167
particle resonance, 131f, 154

parton, 170
photoelectron, 7, 12, 83
pion, 129, 150
plutonium, 110, 114
polarization of particle beams, 43, 49, 64, 127, 130
positron, 83, 144
positronium, 147
potential barrier, 54ff, 180
Powell, Cecil F., 129, 150
proportional counter, 85
proton, 27
psi-particles, 171

Q-value of reaction, 54, 100ff
quality factor; see relative biological effectiveness
quantum electrodynamics, 128, 150
quark, 169

rad, 83
radioactive,
 constant, 18, 53, 57
 series, 20, 21, 52
radioactivity, 15ff
radius of nucleus, 48ff, 54
ranges of particles, 23, 81
reactions, 38, 42, 60, 100ff
reactors, 114ff
relative biological effectiveness, 84
relativity, 11, 30, 72, 76, 144, 166
rem, 84
repulsive core, 129
resonances,
 neutron, 106f, 157
 baryon, 131f, 155, 158
 meson, 131f, 158
roentgen, 15, 85
rotational states, 137
Rutherford, Ernest, 2, 16, 18, 23ff, 27, 38ff, 43ff, 50, 60, 104, 175

scattering, 37ff, 42, 53, 104, 175
scintillation counter, 88
sector focusing cyclotron, 75
semiconductor detector, 87
semi-empirical mass formula, 138ff
shell model, 133f
spark chamber, 92
spectra,
 alpha, 50ff
 beta, 25, 52
spin, 135f, 145, 152f, 159
spin-dependent interaction, 127
spin-orbit interaction, 127, 135f
splitting,
 of energy levels, 135f
 of masses, 163

strangeness, 160
supermultiplet, 163
symmetry, 163, 166ff, 184
synchrocyclotron, 72
synchrotron, 74

tandem (van de Graaff)
accelerator, 68
thermonuclear reaction, 116
bomb, 118
Thomson, Joseph J., 1, 5ff, 27, 39, 93
time reversal, 167
transuranic elements, 94, 107, 110

unified mass unit, 101
uranium, 95, 107, 113

van de Graaff accelerator, 67
vibrational states, 137

wave equation, 54, 121f, 123ff, 133, 144, 180
weak interaction, 130, 159ff, 162
Wigner, Emil, 126, 159
wire chamber, 92

x-rays, 15, 36, 103

Yukawa potential, 128f

Physical constants and conversion factors

Avogadro constant	L or N_A	$6{\cdot}022 \times 10^{23}\ \mathrm{mol}^{-1}$
Bohr magneton	μ_B	$9{\cdot}274 \times 10^{-24}\ \mathrm{J\ T}^{-1}$
Bohr radius	a_0	$5{\cdot}292 \times 10^{-11}\ \mathrm{m}$
Boltzmann constant	k	$1{\cdot}381 \times 10^{-23}\ \mathrm{J\ K}^{-1}$
charge of an electron	e	$-1{\cdot}602 \times 10^{-19}\ \mathrm{C}$
Compton wavelength of electron	$\lambda_C = h/m_e c$	$= 2{\cdot}426 \times 10^{-12}\ \mathrm{m}$
Faraday constant	F	$9{\cdot}649 \times 10^{4}\ \mathrm{C\ mol}^{-1}$
fine structure constant	$\alpha = \mu_0 e^2 c/2h$	$= 7{\cdot}297 \times 10^{-3}\ (\alpha^{-1} = 137{\cdot}0)$
gas constant	R	$8{\cdot}314\ \mathrm{J\ K}^{-1}\ \mathrm{mol}^{-1}$
gravitational constant	G	$6{\cdot}673 \times 10^{-11}\ \mathrm{N\ m}^2\ \mathrm{kg}^{-2}$
nuclear magneton	μ_N	$5{\cdot}051 \times 10^{-27}\ \mathrm{J\ T}^{-1}$
permeability of a vacuum	μ_0	$4\pi \times 10^{-7}\ \mathrm{H\ m}^{-1}$ exactly
permittivity of a vacuum	ϵ_0	$8{\cdot}854 \times 10^{-12}\ \mathrm{F\ m}^{-1}\ (1/4\pi\epsilon_0 = 8{\cdot}988 \times 10^{9}\ \mathrm{m\ F}^{-1})$
Planck constant	h	$6{\cdot}626 \times 10^{-34}\ \mathrm{J\ s}$
(Planck constant)/2π	$\hbar$	$1{\cdot}055 \times 10^{-34}\ \mathrm{J\ s} = 6{\cdot}582 \times 10^{-16}\ \mathrm{eV\ s}$
rest mass of electron	m_e	$9{\cdot}110 \times 10^{-31}\ \mathrm{kg} = 0{\cdot}511\ \mathrm{MeV}/c^2$
rest mass of proton	m_p	$1{\cdot}673 \times 10^{-27}\ \mathrm{kg} = 938{\cdot}3\ \mathrm{MeV}/c^2$
Rydberg constant	$R_\infty = \mu_0^2 m_e e^4 c^3/8h^3$	$= 1{\cdot}097 \times 10^{7}\ \mathrm{m}^{-1}$
speed of light in a vacuum	c	$2{\cdot}998 \times 10^{8}\ \mathrm{m\ s}^{-1}$
Stefan–Boltzmann constant	$\sigma = 2\pi^5 k^4/15h^3c^2$	$= 5{\cdot}670 \times 10^{-8}\ \mathrm{W\ m}^{-2}\ \mathrm{K}^{-4}$
unified atomic mass unit (^{12}C)	u	$1{\cdot}661 \times 10^{-27}\ \mathrm{kg} = 931{\cdot}5\ \mathrm{MeV}/c^2$
wavelength of a 1 eV photon		$1{\cdot}243 \times 10^{-6}\ \mathrm{m}$

1 Å $= 10^{-10}$ m; 1 dyne $= 10^{-5}$ N; 1 gauss (G) $= 10^{-4}$ tesla (T);
0°C $= 273{\cdot}15$ K; 1 curie (Ci) $= 3{\cdot}7 \times 10^{10}\ \mathrm{s}^{-1}$;
1 J $= 10^7$ erg $= 6{\cdot}241 \times 10^{18}$ eV; 1 eV $= 1{\cdot}602 \times 10^{-19}$ J; 1 $\mathrm{cal_{th}} = 4{\cdot}184$ J;
ln 10 $= 2{\cdot}303$; $\ln x = 2{\cdot}303 \log x$; e $= 2{\cdot}718$; log e $= 0{\cdot}4343$; $\pi = 3{\cdot}142$